T0155455

# Kafka Troubleshooting in Production

## Stabilizing Kafka Clusters in the Cloud and On-premises

Elad Eldor

Apress®

***Kafka Troubleshooting in Production: Stabilizing Kafka Clusters in the Cloud and On-premises***

Elad Eldor
Tel Aviv, Israel

ISBN-13 (pbk): 978-1-4842-9489-5
ISBN-13 (electronic): 978-1-4842-9490-1
https://doi.org/10.1007/978-1-4842-9490-1



Managing Director, Apress Media LLC: Welmoed Spahr
Acquisitions Editor: Susan McDermott
Development Editor: Laura Berendson
Editorial Project Manager: Jessica Vakili
Copy Editor: Kezia Endsley

Cover image by Meg (@megindoors) from Unsplash

Distributed to the book trade worldwide by Springer Science+Business Media New York, 1 New York Plaza, Suite 4600, New York, NY 10004-1562, USA. Phone 1-800-SPRINGER, fax (201) 348-4505, e-mail orders-ny@springer-sbm.com, or visit www.springeronline.com. Apress Media, LLC is a California LLC and the sole member (owner) is Springer Science + Business Media Finance Inc (SSBM Finance Inc). SSBM Finance Inc is a **Delaware** corporation.

For information on translations, please e-mail booktranslations@springernature.com; for reprint, paperback, or audio rights, please e-mail bookpermissions@springernature.com.

Apress titles may be purchased in bulk for academic, corporate, or promotional use. eBook versions and licenses are also available for most titles. For more information, reference our Print and eBook Bulk Sales web page at http://www.apress.com/bulk-sales.

Any source code or other supplementary material referenced by the author in this book is available to readers on GitHub. For more detailed information, please visit https://www.apress.com/gp/services/source-code.

Paper in this product is recyclable

*In memory of Gal Eidinger*

# Table of Contents

# About the Author

**Elad Eldor** leads the DataOps group in the Grow division at Unity, formerly ironSource. His role involves preventing and solving stability issues, improving performance, and reducing cloud costs associated with the operation of large-scale Druid, Presto, and Spark clusters on the AWS platform. Additionally, he assists in troubleshooting production issues in Kafka clusters.

Elad has twelve years of experience as a backend software engineer and six years as a Site Reliability Engineer (SRE) and DataOps for large, high throughput, Linux-based big data clusters.

Before joining ironSource, Elad worked as a software engineer and SRE at Cognyte, where he developed big data applications and was in charge of the reliability and scalability of Spark, Presto and Kafka clusters in on-premises production environments. His primary professional interests are in improving performance and reducing cloud costs associated with big data clusters.

# About the Technical Reviewer

**Or Arnon** was born and raised in Tel Aviv, the startup hub of Israel, and, according to him, one of the best cities in the world. As a DevOps engineer, Or thrives on problem-solving via collaboration, tools, and a build-for-scale attitude. He aspires to make good things great. As a manager, Or focuses on business and team growth. He aims to build a team that is challenged, agile, and takes pride in their work.

# Acknowledgments

I'd like to thank the many DevOps, DataOps, and system administrators and developers who maintain Apache Kafka clusters. Your efforts make this book necessary and I hope it will assist you in handling and even preventing production issues in your Kafka clusters.

Special thanks to Evgeny Rachlenko. Our coffee breaks which were filled with discussions about Linux performance tuning sparked my deep interest in this topic, and the knowledge I gained on this topic has been invaluable in my work with Kafka.

Sofie Zilberman encouraged me to focus on JVM tuning and then also introduced me to Kafka. I ended up having these issues as my biggest interests along with Linux performance tuning, and this wouldn't have happened without her. I am indebted to her for setting such a high bar for me during the years we worked together.

Uri Bariach worked with me on troubleshooting dozens of production issues in on-premises Kafka clusters. I'd like to thank him for being such a great colleague and also for editing the on-premises chapters of this book.

I'm grateful to Or Arnon, who works with me at ironSource (now Unity). We spent dozens of hours together analyzing production issues in high throughput, cloud-based Kafka clusters. He is one of the most thorough and professional KafkaOps out there, and his technical editing of this book has been indispensable.

Writing this book was at times daunting. But taking care of Kafka clusters, whether on-premises or cloud-based, is even more challenging. Of all the open-source frameworks I've worked with in production, Kafka is by far the toughest one to handle, and usually the most critical one as well. My hope is that this book will help those who maintain Kafka clusters to reduce the chances of production issues and serve its purpose well.

# Introduction

Operating Apache Kafka in production is no easy feat. As a high-performing, open-source, distributed streaming platform, it includes powerful features. However, these impressive capabilities also introduce significant complexity, especially when deploying Kafka at high scale in production.

If you're a Kafka admin—whether a DevOps, DataOps, SRE, or system administrator—you know all too well the hefty challenges that come with managing Kafka clusters in production. It's a tough task, from unraveling configuration details to troubleshooting hard-to-pinpoint production issues. But don't worry, this book is here to lend a helping hand.

This practical guide provides a clear path through the complex web of Kafka troubleshooting in production. It delivers tried and tested strategies, useful techniques, and insights gained from years of hands-on experience with both on-premises and cloud-based Kafka clusters processing millions of messages per second. The objective? To help you manage, optimize, monitor, and improve the stability of your Kafka clusters, no matter what obstacles you face.

The book delves into several critical areas. One such area is the instability that an imbalance or loss of leaders can bring to your Kafka cluster. It also examines CPU saturation, helping you understand what triggers it, how to spot it, and its potential effects on your cluster's performance.

The book sheds light on other key aspects such as disk utilization and storage usage, including a deep dive into performance metrics and retention policies. It covers the sometimes puzzling topic of data skew, teaching you about Kafka's various skew types and their potential impact on performance. The book also explains how adjusting producer configurations can help you strike a balance between distribution and aggregation ratio.

Additionally, the book discusses the role of RAM in Kafka clusters, including situations where you might need to increase RAM. It tackles common hardware failures usually found in on-premises data centers and guides you on how to deal with different disk configurations like RAID 10 and JBOD, among other Kafka-related issues.

Monitoring, an essential part of any KafkaOps skill set, is addressed in detail. You'll gain a deep understanding of producer and consumer metrics, learning how to read them and what they signify about your cluster's health.

Whether you're a DevOps, a DataOps, a system administrator, or a developer, this book was created with you in mind. It aims to demystify Kafka's behavior in production environments and arm you with the necessary tools and knowledge to overcome Kafka's challenges. My goal is to make your Kafka experience less overwhelming and more rewarding, helping you tap into the full potential of this powerful platform. Let's kick off this exploration.

CHAPTER 1

# Storage Usage in Kafka: Challenges, Strategies, and Best Practices

Kafka's storage intricacies are vital to its operation, and this chapter dives into those details. It begins by exploring how Kafka can run out of available disk space for storing segments. Next, it discusses the challenges of handling data loss due to retention policies and how to manage consumer lag.

The chapter also looks at how to handle sudden data influxes from producers and the importance of understanding daily traffic variations. Compliance with batch durations is examined, followed by guidelines for adding more storage to a Kafka cluster. Finally, the chapter delves into strategies for extended retention.

Throughout the chapter, I break down practical implications, evaluate potential risks, and offer strategies. All of these elements are aimed at optimizing Kafka's storage management.

## How Kafka Runs Out of Disk Space

One of the reasons that a Kafka cluster might halt or stop functioning is that the disks of one or more brokers get filled up. That's why it's important to prevent Kafka brokers from reaching a point at which the segments in their partitions fill up the disks.

There are several scenarios that can cause the disks of a Kafka broker to become full. In order to understand these scenarios, imagine you have the Kafka cluster shown in Table 1-1.




E. Eldor, *Kafka Troubleshooting in Production*, https://doi.org/10.1007/978-1-4842-9490-1_1

**Note** For simplicity reasons, I assume that all the topics in this cluster have the same characteristics:

- Size on disk
- Retention
- Replication factor
- Number of partitions

*Table 1-1.* *An Example of a Kafka Cluster*

| Number of Brokers | 3 | Notes |
|---|---|---|
| Broker Type | i3en.xlarge (running on AWS) | |
| Disk Configuration | JBOD | |
| Disk Size in Each Broker | 2.5TB | |
| Total Disk Size | 7.5TB | |
| Storage to Use | 6.3TB | Assuming you don't use more than 85% of the given storage, to allow 15% of the disk space free |
| Replication Factor | 2 | |
| Number of Topics | 10 | |
| Number of Partitions per Topic | 5 | |
| Retention | 5 hours | |
| Topic Size on Disk | 300GB | The size of the topic, after compression, but before the replication factor |
| Topic Size on disk, After Replication | 600GB | |
| Used Storage (in GB) | 6GB | |
| Used Storage (in %) | 80% | |

Let's look at some of the reasons that Kafka disks can become full:

- *Increasing the replication factor (RF) of existing topics:* Increasing the RF from 2 to 3 will increase the topic size on-disk (after replication) by 300GB (from 600GB to 900GB).
- *Increasing retention for existing topics:* Increasing the retention from 5 to 10 hours will gradually increase the topic size on-disk in the following manner:
  - Before replication: By 300GB, from 300GB to 600GB
  - After replication: By 600GB, from 600GB to 1200GB
- *Adding partitions to existing topics:* In general, adding more partitions to a topic can enhance the concurrency of producers and consumers. For example, if the number of partitions in a topic increases from 90 to 100, this allows producers and consumers to potentially handle more simultaneous operations. This enhancement is tied to the overall capability of handling more parallel tasks in both producing and consuming from that topic.

  However, if a topic has a size-based retention policy and segments are deleted when they reach the size limit, adding more partitions will also increase the total size usage.
  - Before replication: By 60GB, from 300GB to 360GB
  - After replication: By 120GB, from 600GB to 720GB
- *Adding topics*: A new topic will add 600GB of storage on-disk, after replication.
- *Disk failures:* When at least one disk fails on at least one broker in the cluster, and the following conditions apply:
  - The disks are configured in JBOD (instead of RAID).
  - The replication factor is at least 2.

Then:

- The data that resided on the failed disk will need to be replicated to some other disk.
- The total available storage in the cluster is reduced while the used storage remains the same.
- If the storage usage was high enough already, then even a single failed disk can cause the storage to become full, and the broker can fail.

In the previous example, each failed disk will reduce 2.5TB of disk space.

- *Writing to the disks by other processes:* If there are other processes writing to the disks where the log segments are stored, they will reduce the disk space needed to store these segments.

# A Retention Policy Can Cause Data Loss

Data loss happens when some messages aren't read, because they were deleted from Kafka before the consumers had a chance to consume them.

There are several scenarios in which the retention policy can cause data loss for the consumers that consume data from that topic. We'll go over each of them in more detail, but first let's go over on how to configure retention for Kafka topics.

# Configuring a Retention Policy for Kafka Topics

In Kafka, a topic is represented at the disk level as a directory containing child directories, with each child directory representing a partition of that topic. These topics are housed under a directory specified by the `log.dirs` configuration parameter. The data being written to the topic is stored in log segment files within the partition directories. To avoid a situation where the directory that contains all the topics becomes full, two configuration parameters control the retention of the data in the topic: `log.retention.bytes` and `log.retention.hours`. These settings help manage storage and ensure that old data can be purged in order to make room for new messages:

- `log.retention.bytes`: This configuration sets the maximum size of all log segments inside a partition. Once that size is reached, the oldest log segments are deleted in order to free up space. This configuration applies per partition.
- `log.retention.hours`: This configuration sets the maximum number of hours a message will be retained in the partition before being deleted. It applies on a per-topic basis. Unlike the previous byte-based retention policy, this time-based policy refers to individual messages rather than segments. A segment is considered to have passed its retention period only when all the messages in that segment have exceeded their retention period.

In order to learn how these retention policies control the storage usage inside a topic directory, imagine a topic with the characteristics shown in Table 1-2.

***Table 1-2.*** *An Example of a Kafka Topic*

| Number of Partitions | 50 |
|---|---|
| log.retention.bytes | 10GB |
| log.retention.hours | 1 |
| Message Size on Disk (After Compression) | 1KB |
| Avg Produce Rate | 10K/sec |
| Segment Size (Determined by log.segment.bytes) | 1GB |
| Produce Rate | 1K/sec |
| Replication Factor | 2 |

In this case, segments will be deleted from the partition either when all the messages inside these segments are older than one hour or when the size of all the segments inside the partition reaches 10GB, whatever comes first.

After configuring this topic, let's consider the chances of losing data.

## Managing Consumer Lag and Preventing Data Loss

When consumers fall into lag, some of them can consume older messages at any given time. If this lag continues to grow and surpasses the threshold of one hour (as specified by the `log.retention.hours` value in the example), those consumers may attempt to read messages that have already been deleted due to the retention policy. When this occurs, these consumers lose all the deleted data, and if they are low-level consumers (like Apache Druid, Spark Streaming, or Spark Structured Streaming applications that consume from Kafka), they might crash. This crash happens because they try to consume a message from an offset that no longer exists in the partition they consume from. This situation might lead to downtime for the consumers between the time they crash until they restart. After restarting, they will only read messages from the oldest offset that exists in the partition.

To prevent this scenario, you should monitor the consumers' lag and correlate between consumers and the topics they consume from. If the total time of the consumer lag approaches the topic retention, it's essential to raise an alert.

To mitigate this scenario, you can increase the retention of the topic if its consumers are lagging. If the lagging application has fewer consumers than the number of partitions in the topic it consumes from, adding more consumers is advised. However, it's crucial to distribute consumers evenly among the number of partitions. This doesn't mean that the number of consumers and partitions must be equal, but rather that the number of partitions should be divisible without a remainder by the number of consumers in each consumer group.

## Handling Bursty Data Influx from Producers

Sometimes, producers may temporarily inject much more data into a Kafka topic. This can be due to a configuration change or because a sudden burst of data has arrived at the producer, such as when a producer recovers from downtime and writes all accumulated data at once.

In such cases, the partition's configured size limit (`log.retention.bytes`, 10GB per partition in the current example) can be reached more quickly than usual. Consequently, the time that segments remain in the partition will be shorter. If there are any consumers lagging behind at that moment, this can lead to data loss.

To monitor such an issue, keep an eye on sudden increases in the producing rate into the topic, as this might signal an abnormal data influx. If the topic starts receiving more data, consider increasing its retention in order to give lagging consumers more time to catch up. If the lagging application has fewer consumers than the number of partitions in the topic it reads from, consider adding more consumers. Doing so can help distribute the load more evenly and prevent data loss.

## Balancing Consumer Throttling and Avoiding Unintended Lag

When consumers begin to lag behind (due to reasons such as an input rate higher than the processing rate, a sudden spike in traffic, a restart of the consumer, etc.), it may be wise to regulate the amount of data each consumer reads. This can be done by using the `max.partition.fetch.bytes` configuration, which limits the amount of data consumed by a consumer in each batch.

While this throttling can contribute to the stability of consumers, it may lead to additional lag if the consumption rate falls below the input rate into the topic. This can be a delicate balance to strike. In the hypothetical scenario, if the backlog of unread messages grows to an hour and the consumer subsequently attempts to read from an offset that's older than one hour, it may fail to find the required data. This failure would occur because the messages at that offset would have already been deleted due to the topic's retention policy.

## Understanding Daily Traffic Variations and Their Impact on Data Retention

In the previous example, the average input rate into the topic is 10K/sec, and the processing rate of consumers is 10K/sec. On paper, this means that consumers won't get into lag, and unless they crash, it's hard to see how they can lose data. However, if there's a big variance of traffic during the day, there's still a chance that consumers will have a backlog and might also lose data.

For example, if the input rate to the topic behaves as following:

- Low hours between 12 am-8 am, it's 2K/sec
- Medium hours between 9am-5pm, it's 5K/sec
- Peak hours between 7pm-11pm, it's 18K/sec

then during the peak hours, the consumers might get into lag.

If, after peak time is over, some segments (that the consumers didn't manage to consume) still exist in Kafka, then the consumer will manage to read all of these events. However, if for some reason, the consumers developed a lag of more than 60 minutes during peak time, there's a chance that once peak time is over, the consumers will develop a delay that's longer than the 60 minutes, and in such cases the consumers will lose that data.

## Ensuring Batch Duration Compliance with Topic Retention to Avoid Data Loss

Non real-time consumers often operate on a batch duration, which refers to the regular intervals at which they wake up and consume messages from the topic partitions. If the retention time for the topic they are reading from is, for example, 60 minutes, this can be risky. Once the consumers begin reading from the topic, older segments in the partitions may be deleted. If this happens, consumers may lose the data in those partitions, or even encounter errors because they are attempting to read from an offset that has been deleted. To avoid such problems, it's vital to consider the topic's retention time when configuring the batch interval for non-real-time consumers. Specifically, the retention period should be longer than the batch duration to ensure that no data is lost.

# Adding Storage to Kafka Clusters

When looking to provision a Kafka cluster with additional disk space, you generally have a couple of options: you can attach new disks to each of the existing brokers or add more brokers to the cluster. The decision between these options depends on various factors related to your specific setup and needs, and understanding the implications of each choice is key to making the right decision. The following sections delve into these options and explain when and why you might choose one over the other.

## When the Cluster Is On-Prem

When you need to add more disk space to an on-premises Kafka cluster, there are various strategies to consider, each addressing different scenarios and needs. The primary options include scaling up the existing brokers or scaling out by adding new brokers to the cluster.

## Scaling Up

Scaling up refers to expanding the storage in the existing brokers. This can be done in several ways:

- *Adding a drawer to a broker without drawers:* For brokers that initially only use built-in disks and have no external drawers, adding a drawer can be a viable option if the broker configuration allows it. This involves selecting and attaching a compatible drawer, then migrating the data as necessary. This approach can provide a substantial increase in storage without the need to replace or modify the existing built-in disks.
- *Replacing drawers with larger ones:* If the existing disk drawer attached to the brokers is filled, you can replace it with a larger drawer that can accommodate more disks. The process involves substituting the existing drawer with a new, bigger one across all the brokers, connecting the new drawers, migrating the disks from the old drawers to the new ones, and adding additional disks if needed.
- *Adding more drawers to the brokers:* Alternatively, you may choose to attach more drawers to each broker, thus expanding the storage capacity. This option is available when the broker configuration supports additional drawer connections. The same migration process applies, moving Kafka's data directories to the new disks and ensuring that the existing data remains accessible.

### Scaling Out

Scaling out refers to the addition of new brokers to the cluster. This option helps when you need to substantially increase storage capacity. Adding more brokers inherently means adding more disks and storage to the cluster. Depending on the configuration and needs, this can be a viable option to manage a significant growth in data.

In all cases, care must be taken to properly manage the migration of data. When moving Kafka's data directories to new disks, you'll need to mount the appropriate directories on to the new disks and transfer all necessary data. This ensures that the existing data remains accessible to the Kafka brokers, effectively expanding the storage without risking data loss or unnecessary downtime.

## When the Cluster Is in the Cloud

The strategy for expanding storage in Kafka brokers in the AWS cloud without adding more brokers to the cluster differs depending on the type of disks used. The following sections explain the use of EBS and NVME disks.

### EBS Disks

If the brokers contain only EBS disks, there are two main options for increasing storage capacity:

- *Scale up:* This involves increasing the existing EBS size or attaching more EBS disks to the existing brokers. Since EBS disks can be resized without losing data, this process can be accomplished with minimal disruption. However, it may require stopping the associated instances temporarily to apply the changes, while maintaining data consistency throughout.
- *Scale out:* Scaling out is achieved by adding more brokers, each with their own EBS disks. This effectively increases the total storage capacity across the cluster. As new brokers are added, it's crucial to ensure they are integrated correctly into the cluster without impacting existing data and performance.

## NVME Disks (Ephemeral Disks)

When the brokers contain ephemeral disks, such as NVME disks in AWS, scaling up or scaling out the Kafka cluster becomes possible. However, these options must be handled with care:

- *Scale up:* Scaling up keeps the same number of brokers in the cluster, but each broker will have more disks than before. The process typically involves stopping and starting a broker, which can cause data loss on that NVME device since a new one is assigned. To prevent this, a proper migration process must be followed. Data must be moved from the old NVME device to a persistent storage before the broker is stopped. Once the broker is started with the new NVME device, the data can be moved back.
- *Scale out:* Scaling out adds more brokers with the same number of disks as before. This can be done without the same risk of data loss as scaling up, but careful planning and coordination are required to ensure that the new brokers are correctly integrated into the cluster without impacting existing data and performance.

# Strategies and Considerations for Extended Retention in Kafka Clusters

In a Kafka cluster, retention policies need to balance various considerations, including operational recovery, business needs, and data replay requirements. Operational recovery considerations revolve around the time it takes for KafkaOps or development teams to manage consumers and producers. Whether recovering from consumer downtime or dealing with issues in the Kafka cluster, typical retention periods span around ±7-10 hours.

Business needs might also call for more extended retention periods. Customers with on-premises Kafka deployments without immediate KafkaOps support might need days of retention to account for logistical challenges, such as remote support and international communication. These needs become even more pronounced if troubleshooting becomes a prolonged process.

Data replay introduces another dimension to retention. If something in the data was incorrect or corrupted after it was consumed from the Kafka topic, extended retention ensures that the data remains available to correct mistakes or adapt to new requirements. This necessity to replay data in specific scenarios further emphasizes the importance of having a thoughtfully planned retention policy.

You can manage extended retention in a Kafka cluster using several strategies. One method is to increase the cluster's storage, as previously discussed. Another option is to create a custom streaming application that consumes data from topics needing lengthy retention and archives them in deep storage, such as AWS S3. This his allows consumers to retrieve old data when needed.

A promising alternative under development is Kafka's Tiered Storage feature, which creates a two-tier system, using existing broker disks for local storage and systems like HDFS or S3 for remote storage of older log segments. This approach not only enables Kafka to function as a long-term storage solution, but also scales storage independent of memory and CPUs, thus reducing local retention time and removing the need for separate data pipelines to external locations.

# Calculating Storage Capacity Based on Time-Based Retention

When provisioning a Kafka cluster, one of the considerations that you need to consider is the required storage. There are several factors that need to be considered when calculating how much storage the Kafka cluster needs. All the factors are related to topics, so for each topic, you need to know the following:

- Message size after compression
- Number of events/sec during the hours of the day
- Required retention
- Replication factor

The amount of storage used by a topic can then be calculated using this formula:

> Topic_size_on_disk = [average message size] X [average number of messages per sec] X [number of required retention hours] X [replication factor of the topic]/[compression rate]

After performing this calculation for all the topics in the cluster, you get a number that represents the minimal required storage for the cluster. Since it's hard to predict the compression rate, you'll need to test for yourself to see the difference between the ingested data into Kafka and the data that's actually stored in Kafka. That's regardless of whether the compression is performed by the producer or by the Kafka brokers.

But that number is only a starting point, because we also need to add some extra storage due to the following reasons:

- We don't want the used storage to reach above ±85 percent storage use. This threshold should be determined by the KafkaOps team, since there's no right or wrong value.
- The storage requirements may change over time - e.g. the replication factor or the retention might increase

As an example, say you have five topics, as shown in Table 1-3.

***Table 1-3.*** *An Example of Five Topics*

| Topic | Average Message Size (in KB) | Average Number of Messages per Sec | Number of Required Retention Hours | Replication Factor | Compression Rate | Topic Size in GB (Before Compression | Topic Size in TB (After Compression) |
|---|---|---|---|---|---|---|---|
| Topic-1 | 0.5 | 100,000 | 7 | 2 | 80% | 2.3 | 1.9 |
| Topic-2 | 0.7 | 200,000 | 7 | 2 | 70% | 6.6 | 4.6 |
| Topic-3 | 0.4 | 200,000 | 5 | 2 | 60% | 2.7 | 1.6 |
| Topic-4 | 1 | 250,000 | 10 | 2 | 70% | 16.8 | 11.7 |
| Topic-5 | 1.2 | 150,000 | 8 | 2 | 80% | 9.7 | 7.7 |
| Total Storage of All Topics | | | | | | 38.0 | **27.5** |

In this case, the minimal required storage for the cluster is 27.5TB. On top of that, you need to add some storage so you won't use more than 80 percent, so multiply that number by 1.25; you get 34TB.

To conclude, the cluster requires 34TB in order to store the required retention.

## Retention Monitoring

There are certain alerts that should pop up in cases that could lead to the storage getting full in one of the brokers, or data loss for one of the consumers. The next sections discuss them.

## Data Skew in Partitions

If one partition gets more data than the others (this can occur when the producer uses a partition key that isn't optimized for balancing the data among the partitions), and `log.retention.bytes` is configured for its topic, in effect this partition has a lower retention (in terms of partition size) than the other partitions in that topic.

If that data skew is big enough, and the consumer of that partition lags, then it could get to a situation where all the data inside that partition is deleted.

Figure 1-1 shows an example of monitoring the data skew within the partitions of a specific topic.

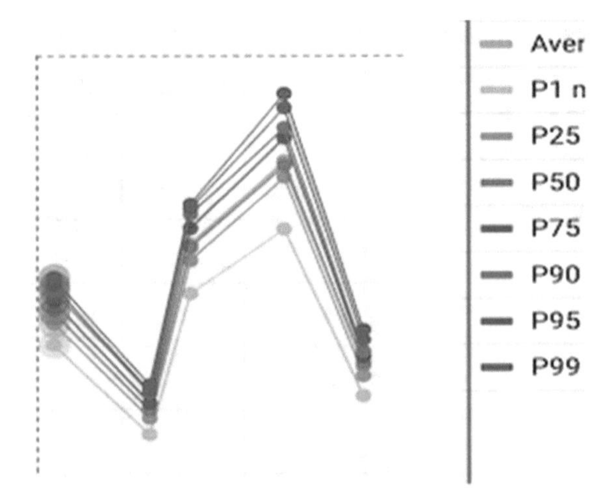


***Figure 1-1.*** *An example of data skew monitoring for a specific topic*

The X axis is time, and the Y axis is the message rate per minute into each partition. Each line represents a partition, and the partitions in this case are sorted by percentiles according to the message rate that is produced into them. So the partition named P1 is the partition that gets the lowest number of messages, P50 is the median, and P99 is the partition that gets the most messages.

## Message Rate Into Topic

If the message rate into a specific topic increases and you want to prevent consumers from losing data, you have the following options:

- Adjust the retention policy of the topic
- Increase the consumption rate of the consumers from that topic
    - If the topic increases by size but only the `log.retention.hours` is configured, this can cause the storage to become full.
    - If both time and based-retention policies are configured and the topic size increases, and if there's a consumer that relies on the retention in order to read data from the topic, it could get into a situation where the data doesn't exist since it was deleted.

It's important to detect an increase in the message rate upfront and then adjust the retention policies accordingly.

## Don't Write to the / Mount Point

In certain situations, the / mount point of one or more brokers may become full, causing the broker to halt. This problem manifests in various ways, including the operating system's reduced functionality or even failure, consumers experiencing difficulties in consuming messages, producers encountering "queue full" errors due to full buffers, and the broker process falling into an uninterruptible state, indicated by a "D" in the state column of the top command.

This issue primarily stems from the / mount point filling up in one or more brokers, leading to the symptoms described previously.

There are a few different scenarios that can lead to this problem. If ZooKeeper (ZK) runs on the Kafka broker machine without retention on ZK backup files or if the purger doesn't work, the backup files can fill the / mount point. Also, a user placing a large file like a log in a directory mounted to / or writing segments to the / directory can cause the same issue. To prevent these occurrences, it's wise to take precautions such as not writing to / or any directory mounted to it, avoiding configuring ZK to write its backup files to /, and not deploying ZK on the Kafka broker's machines.

Careful monitoring is essential to detecting and mitigating this issue before it becomes critical. Notably, the load average may increase when / is full, as applications may become stuck trying to access files from directories mounted to `/`. Close monitoring of available disk space in / and watching consumer lag and producer queue full metrics can also provide early warning signs.

## Summary

This chapter delved into the critical role that disk storage plays in Kafka clusters, illuminating various aspects of Kafka's storage usage. It began by exploring how the disks of Kafka brokers can become filled, causing the cluster to halt. Various scenarios were explored, from increasing the replication factor and retention of topics to disk failures and unintended writes by other processes.

The chapter further delved into the importance of retention policies in Kafka, exploring how they can cause data loss for consumers and the critical configuration parameters needed to avoid such situations. Several scenarios were analyzed, including consumer lag management, handling unexpected data influx, balancing consumer throttling, understanding traffic variations, and ensuring batch duration compliance.

Strategies for adding more storage to a Kafka cluster were also explored, focusing on different approaches depending on the cluster's location (on-prem or in the cloud). The chapter concluded with an overview of considerations for extended retention in Kafka clusters and guidelines for calculating storage capacity based on time-based retention.

Overall, this chapter aimed to provide you with a comprehensive overview of storage usage in Kafka clusters and help you avoid common storage-related pitfalls.

The next chapter explores a range of producer adjustments, from partitioning strategies that balance even distribution and clustering of related messages, to the fine-tuning of parameters like `linger.ms` and `batch.size` for enhanced speed and response time. We'll also examine how data cardinality influences Kafka's performance and uncover scenarios where duplicating data across multiple topics can be an intelligent strategy.

CHAPTER 2

# Strategies for Aggregation, Data Cardinality, and Batching

This chapter digs into various adjustments you can make to the Kafka producer that can notably increase your Kafka system's speed, response time, and efficiency. The chapter starts by exploring the partitioning strategy, which aims for an equilibrium between distributing messages evenly and clustering related messages together. It will then dive into adjusting parameters like `linger.ms` and `batch.size` to improve speed and decrease response time. From there, we'll learn how the uniqueness and spread of data values, known as *data cardinality,* impact Kafka's performance. And finally, we'll explore why, in some cases, duplicating data for different consumers can be a smart move.

## Balancing Message Distribution and Aggregation for Optimal Kafka Performance

The key to effective partitioning in Kafka revolves around finding the balance between equal message distribution and effective aggregation. An understanding of your consumers' processing requirements is crucial in making this decision.

A popular and straightforward approach that is often implemented by producers is the round-robin method, which ensures an equal number of messages is sent to each partition. When your consumers don't require message aggregation, this method is highly effective. It promotes stability of the Kafka brokers as each broker handles a balanced load of messages. Moreover, it simplifies message consumption when each message is handled independently.




E. Eldor, *Kafka Troubleshooting in Production*, https://doi.org/10.1007/978-1-4842-9490-1_2

However, for certain consumers, such as Apache Druid or other stream-processing systems like Apache Flink or Apache Samza, this method might not be ideal. These systems often need to aggregate incoming messages, and for that, a round-robin strategy can lead to inefficient processing. Similar messages may end up being spread across different partitions, making aggregation more complicated and less efficient.

When dealing with consumers that require aggregation, it's beneficial to use a partition key that can hash some fields of the message's value. All messages with the same key will then be produced to the same partition, leading to similar messages being grouped together. This approach, however, might lead to uneven message distribution across partitions, as some partitions may end up with more messages than others.

When the consumer is a stream-processing system or a database that benefits from related data being grouped together, using a partition key becomes important. The partition key can ensure data relatedness—for instance, all clicks from a particular user session in a clickstream data scenario can end up in the same partition, making real-time sessionization simpler and more efficient.

In the end, the choice of partitioning strategy—whether round-robin or using a partition key—boils down to the nature of your consumers and their processing requirements. The challenge is to find the sweet spot between two key needs: grouping related messages together for efficient aggregation and ensuring an even distribution of messages across all partitions to maintain Kafka's performance and stability.

# Tuning Parameters to Increase Throughput and Reduce Latency

`linger.ms` and `batch.size` are two influential parameters that profoundly impact the behavior and performance of producers, Kafka brokers, and consumers.

When configuring the `linger.ms` parameter, a producer can intentionally delay sending messages by grouping them into larger batches if batching is enabled, or simply hold them for a specified time if batching is disabled. This waiting time is counted in milliseconds. On the other hand, the `batch.size` parameter allows the producer to manage the maximum size of these batches in bytes. These settings represent the delicate balance between latency and throughput, fundamental tradeoffs in messaging systems.

## Optimizing Producer and Broker Performance: The Impact of Tuning linger.ms and batch.size in Kafka

The producer is the primary component that's affected by changes in the `linger.ms` and `batch.size` parameters. By extending `linger.ms`, a producer can group more messages together in a batch (if batching is enabled) before dispatching it to Kafka, thereby lowering the number of network requests and increasing throughput. On the flip side, longer lingering can increase latency, which might cause delays in applications that require real-time responses.

Furthermore, a larger `batch.size` offers multiple benefits. It not only reduces network requests but also enables better compression, thereby making the data transfer more efficient. This improvement in compression can lead to reduced storage requirements and improved overall performance. However, larger batches also impact the producer's buffer, occupying more buffer space. If the buffer fills up faster than messages can be sent, the producer might have to wait, affecting performance negatively.

At the Kafka brokers, tuning the `linger.ms` and `batch.size` configurations can lead to significant performance improvements. Larger batches from the producer translate to fewer network requests for the broker to manage. This decrease in network requests conserves the CPU, which can be utilized elsewhere.

Moreover, when it comes to disk storage, larger batches provide a dual benefit. Firstly, they enhance the efficiency of message compression (assuming it was performed by the producer) within the batch, allowing for more effective storage and potentially reducing disk I/O. Secondly, they improve disk utilization by filling disk blocks more efficiently.

However, these larger batches also affect the broker's RAM use. Since Kafka utilizes the OS page cache for caching messages, larger batches can take up more space in the page cache, potentially reducing the available cache for messages arriving from other producers or other data.

## Understanding Compression Rate

When it comes to compression, it is essential to note that the process takes place at the producer before the messages are sent. The size of the batches significantly influences compression rates. For instance, a batch of 100 messages might achieve a compression rate of 2x, while a larger batch of 1,000 similar messages might reach a compression rate of 10x. The key factor here is the number of similar messages in the batch, which when larger, provides a better scope for compression, reducing network and storage use.

Lastly, the way you partition your data can also influence compression rates. For example, when using a SinglePartitionPartitioner, where the producer writes only to a single, randomly selected partition, larger batches can enhance compression ratios by up to two times. This is because all messages in a batch belong to the same partition, and the likelihood of having similar messages in the batch increases, leading to better compression.

In conclusion, `linger.ms` and `batch.size` are both powerful knobs to tweak in your Kafka setup, allowing you to optimize throughput, latency, and resource use across producers, brokers, and consumers. Still, it is crucial to align these parameters with your specific use case so you avoid potential bottlenecks or inefficiencies that can emerge from misconfiguration.

# The Effect of Data Cardinality on Producers, Consumers, and Brokers

*Data cardinality* refers to the uniqueness and distribution of values in a dataset, and it has a significant impact on the operations and performance of Kafka's ecosystem. This influence is felt across the producers, brokers, and consumers in different ways, from compression and storage handling to processing and querying information. This section explores the concept of data cardinality, detailing its effects on different Kafka components, and outlines practical ways to control and optimize it for improved performance and efficiency in Kafka operations.

## Defining Data Cardinality

*Data cardinality,* a concept referring to both the quantity of unique values (*cardinality value*) and their spread across a field (*cardinality distribution*), plays a pivotal role in the overall performance of Kafka's producers, brokers, and consumers.

First, I explain the terms *cardinality value* and *cardinality distribution* using an example of an e-commerce website that logs user activities. One of the fields in the log data might be `activity_type`, which could have values like `viewed_product`, `added_to_cart`, `checked_out`, and so on.

In this case, the *cardinality level* would be the total number of unique activity types. The *cardinality distribution*, on the other hand, shows the frequency or proportion of each activity type in the overall data. For instance, you might find that 70 percent of the activities are `viewed_product`, 20 percent are `added_to_cart`, and 10 percent are `checked_out`. This distribution can significantly impact the processing and analysis of the data, especially when dealing with operations like compression and aggregation.

## Effects of High Data Cardinality

The interaction of data cardinality with Kafka components directly affects compression in both Kafka producers and brokers. High cardinality often results in a lower compression ratio because the diverse data has fewer repeated elements that compression algorithms can capitalize on. This decrease in compression ratio not only leads to larger batch sizes but also increases network bandwidth utilization, due to the transmission of these bigger messages. This rise in message sizes also has a downstream effect on Kafka brokers, demanding more disk space for storage.

At the same time, the CPU requirements may escalate on the producer or broker side if a CPU-intensive compression algorithm is employed. The additional computational demand is due to the increased complexity and reduced efficiency of compressing high-cardinality data.

Furthermore, high data cardinality influences the performance of consumers that rely heavily on aggregation, such as Apache Spark Streaming, Apache Druid, Apache Flink, and Apache Samza. The aggregation ratio generally decreases as data cardinality increases, which can impact the throughput and overall efficiency of these consumers.

Beyond compression and aggregation, memory utilization in Kafka brokers and consumers can be affected by data cardinality. Particularly for consumer applications that employ in-memory structures like hashmaps to store unique keys, high cardinality implies a larger number of unique keys, thereby demanding more memory.

Moreover, high cardinality data may impact the performance of Kafka consumers that support querying—queries involving fields with high cardinality can be significantly slower due to the large amount of unique data points to process.

To wrap up, data cardinality deeply affects various aspects of Kafka's performance, including but not limited to compression, aggregation, memory use, network utilization, and query performance. Hence, understanding your data and managing cardinality levels appropriately can lead to more efficient Kafka operations.

## Reducing Cardinality Level and Distribution

In many data-processing scenarios, it becomes crucial to mitigate the high cardinality level or distribution of data for more efficient and manageable operations. High cardinality, characterized by a large number of unique values in a field, or an evenly distributed cardinality, where values appear with roughly equal frequencies, can pose challenges for data processing and analytics. This is due to increased memory use, network bandwidth consumption, and a more complex computational requirement. Consequently, devising strategies to reduce the cardinality level and distribution is a significant aspect of data optimization.

Imagine a situation where the data exhibits a high cardinality level but a low cardinality distribution. This is often the case when a field contains numerous unique values, but a handful of these values occur frequently, while the majority are infrequent. A practical approach to address this scenario involves *bucketing* or grouping infrequent values together. For instance, if you deal with data containing IP addresses and find that many of them occur sporadically, you could aggregate these rare IP addresses into a generic category like `Other_IPs`. This tactic effectively reduces the cardinality level without substantially altering the cardinality distribution.

On the other hand, the data might present both a high cardinality level and high cardinality distribution. This situation signifies that you have a vast range of unique values that are almost uniformly distributed. This poses a greater challenge. However, the problem can be tackled using *bucketing*, where values are grouped based on predetermined rules or ranges. Let's consider timestamps as an example. Instead of using exact timestamps, which naturally results in high cardinality, you could use time ranges (like morning, afternoon, and evening) or even specific dates. This strategy reduces the number of unique values, thereby reducing the cardinality level while still preserving meaningful information for analysis.

By applying these strategic measures, you can effectively control and manage the level and distribution of data cardinality, thereby optimizing your data operations for improved performance and efficiency.

## Duplicating Data to Reduce Latency

In scenarios where multiple consumers read from the same Kafka topic, and the relevance of the data differs among these consumers, it's worthwhile for the producer to write to two distinct Kafka topics—one with all the data and the other with a specific subset of the data.

This approach comes into play when some consumers require the full range of data while others need only a subset. Instead of forcing those consumers who require only part of the data to sift through all the messages in a topic, a separate topic containing only the relevant subset of data can be created.

Separating the data into different topics can enhance processing speed, particularly for consumers that deal with real-time or near-real-time data ingestion. Without the overhead of filtering unnecessary data, these consumers can focus on the core task of processing and storing the relevant data, significantly improving overall performance.

In summary, by producing to two Kafka topics—one containing all the data and the other housing only a specific subset—the data pipeline becomes more efficient and tailored to the needs of different consumers. This reduces unnecessary resource consumption and optimizes processing speed, thereby enhancing the overall performance of the data processing system.

## Summary

This chapter first tackled the task of dividing data across Kafka partitions. The goal is to distribute data uniformly, yet also organize it strategically to meet user's requirements.

Subsequently, the chapter explored the process of adjusting the `linger.ms` and `batch.size` settings. By fine-tuning these configurations, you can utilize Kafka broker's disk space more efficiently, alter the data compression rate, and strike a balance between data transmission speed and volume.

Additionally, the chapter delved into the concept of data cardinality, which is related to the distribution and quantity of unique values in a field. We discussed why an abundance of unique data can negatively impact compression, consume excessive memory, and decelerate queries. As a result, we learned some methods to effectively handle this issue.

Lastly, we learned that duplicating data across multiple topics can sometimes be beneficial. By establishing separate topics for distinct data subsets, we can see an improvement in processing speed for certain consumers. This tactic allows you to enhance the overall efficiency of your data processing system.

The following chapter navigates the intricate landscape of partition skew within Kafka clusters. From the subtleties of leader and follower skews to the far-reaching impacts on system balance and efficiency, the chapter explores what happens when brokers are unevenly loaded. This in-depth examination covers the consequences on production, replication, consumption, and even storage imbalances. Along with a detailed exploration of these challenges, I offer guidance on how to monitor and mitigate these issues, illuminated by real-world case studies.

## CHAPTER 3

# Understanding and Addressing Partition Skew in Kafka

When partitions, be they leaders or followers, are distributed unevenly among brokers within a Kafka cluster, the equilibrium and efficiency of the cluster, as well as its consumers and producers, can be at risk. This inequality of distribution is called *partition skew.*

Partition skew leads to a corresponding imbalance in the number of production and consumption requests. A broker that hosts more partition leaders will inevitably serve more of these requests since the producers and consumers interact directly with the partition leaders.

This domino effect manifests in various ways. Replication on brokers hosting a higher number of partition leaders might become sluggish. At the same time, consumers connected to these brokers may experience delays. Producers might find it difficult to maintain the necessary pace for topic production on these specific brokers. Additionally, there might be a surge in page cache misses on these brokers, driven by increased cache eviction as a result of the higher traffic volume.

These symptoms illustrate how swiftly a mere partition skew can escalate into substantial production challenges within a Kafka cluster. In an ideal setting, all the topic partitions would be uniformly distributed across every broker in the Kafka cluster. Reality, however, often falls short of this perfection, with partition imbalance being a common occurrence.

This chapter is devoted to exploring various scenarios that can lead to such imbalances, providing insights into how to monitor and address these issues.




E. Eldor, *Kafka Troubleshooting in Production*, https://doi.org/10.1007/978-1-4842-9490-1_3

Note that for the sake of clarity, throughout this chapter, the term *partition leader* describes a partition to which producers supply data and from which consumers retrieve it. This term can also describe the partition for which the hosting broker serves as its leader. Utilizing the term *leaders* for partitions rather than associating brokers with leadership of partitions offers a more intuitive understanding of the concept.

# Skew of Partition Leaders vs. Skew of Partition Followers

In a Kafka cluster, partitions come in two forms: leaders and followers. When examining partition skew, it's essential to specify which type of skew you're referring to—whether it's a skew of leaders, followers, or both.

A *leader skew* occurs when a broker holds more partition leaders across all topics in the Kafka cluster. This means that it deals with more producer writes and consumer reads for these topics compared to other brokers that host fewer leaders across the topics.

A *follower skew* takes place when a broker holds more partition followers across the entire range of topics. In this scenario, the broker carries out more replication operations to keep the replicas in sync, regardless of which specific topic they belong to.

# Potential Problems with Brokers That Host Many Partition Leaders

When a broker hosts more partition leaders than other brokers, you first need to map the leaders to the topics they belong to. Once you find the topics, you need to go over them one by one and check for the issues discussed in the following sections.

## Message Rate (or Incoming Bytes Rate)

If a broker hosts more partition leaders of a specific topic than other brokers, and if that topic has a much higher message rate compared to other topics on that cluster, the resulting skew can lead to several challenges. This broker may experience increased disk

I/O operations, particularly if the page cache is unable to fulfill read requests. This would necessitate direct reads from the disk, which can lead to more disk IOPS (input/output operations per second).

Additionally, consumers attempting to read from leader partitions on the skewed broker may face delays, thus affecting the overall data processing times. The situation could also impact producers, who might find that they have more messages in their buffers when writing to the partition leaders on that broker. These delays in transmitting messages, if not managed appropriately, can even lead to data loss.

## Number of Consumers Consuming the Topic

The number of consumers that consume from a topic can greatly affect a broker's behavior, regardless of whether the message rate into that topic is low. When a broker hosts more partition leaders than others, it's more likely to experience issues like high disk I/O reads due to page cache misses. This, in turn, may cause consumers connected to that broker to lag and producers to generate fewer messages than needed because of the strain on the broker's disks.

Therefore, if a particular topic has more consumers compared to others, it makes sense to first determine whether the number of consumers can be trimmed. Sometimes, there might be consumers that are no longer required, like those that are no longer relevant and developers or devops just forgot to shut them down. In other cases, the existing consumer count may simply exceed what's necessary. An example is when an application unnecessarily creates 100 consumers for a topic when only 50 are needed to prevent consumer lag. By identifying and addressing these situations, the broker's operation can be streamlined and optimized, ensuring efficient performance even with a low message rate.

## Number of Producers Producing to the Topic

If the number of producers that produce to this topic is higher than other topics, then the broker that hosts more partition leaders of this topic can suffer from higher context switches and possibly a higher CPU utilization (particularly the `sy%` and `us%` CPU metrics).

## Follower (Replica) Skew in the Broker

An uneven distribution of partition leaders among brokers can lead to a skew of followers as well. Consider the following cluster:

- Three brokers: B1, B2, and B3
- Two topics: T1 and T2
- Both topics have 100 partitions
- The replication factor is 3

Figure 3-1 shows a possible scenario in which there are more partition leaders on broker B1 compared to the other two brokers.

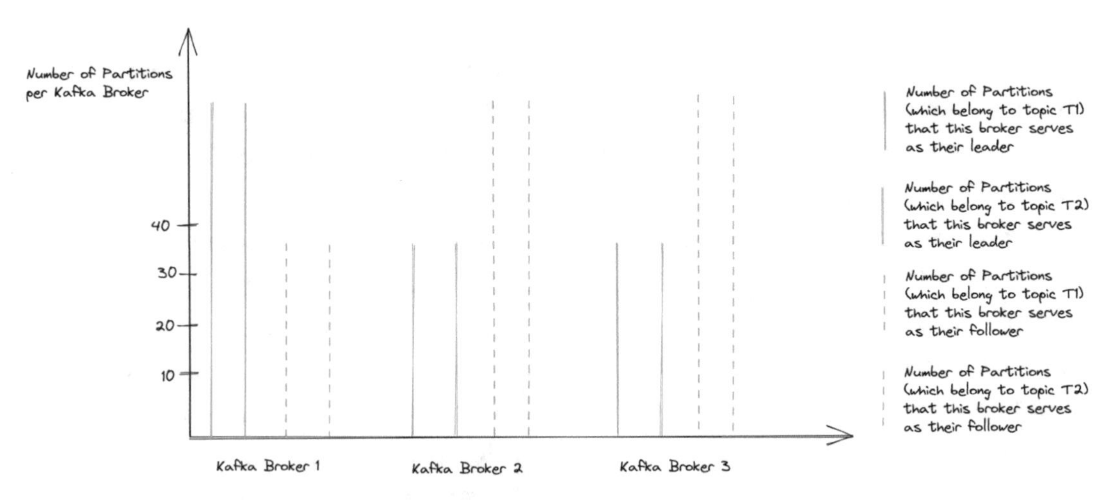


***Figure 3-1.*** *Number of partitions per broker. While broker B1 hosts more partition leaders than brokers B2 and B3, these brokers host more partition followers than broker B1*

In this case, all brokers host the same number of leaders and followers, but B1 holds more leaders, while B2 and B3 host more followers. So a broker with more partition leaders will cause the other brokers to host more partition followers.

This will lead to more fetch requests from the brokers (hosting the partition leaders replicas) into the broker that hosts these leaders (which is also the broker that hosts more leaders). A broker that deals with more fetch requests from other brokers might not be able to serve fetch requests from consumers at the same rate as other, non-skewed brokers.

## Number of In-Sync Replicas in the Broker

A broker that hosts more partition leaders than its counterparts can find itself lagging in the replication of the follower partitions it also hosts, which means this broker might have less in-sync replicas than other brokers.

This happens because hosting more leaders means the broker has to handle an increased number of writes from producers and reads from consumers, adding substantial load to its operations.

Along with managing the leader partitions, the broker is tasked with replicating data for the follower partitions. This requires reading data from the leaders and writing it to the followers. This dual responsibility increases the broker's workload significantly.

This higher load can create competition for the broker's resources, including CPU, memory, network bandwidth, and disk I/O. Striking a balance between the tasks of serving reads and writes for the leader partitions and replicating data for the follower partitions becomes a complex challenge.

The struggle for resources and the heightened workload may cause the broker to fall behind in replicating data to the follower partitions. The process can slow down, particularly if the resources are mostly consumed in handling the leader partitions.

Such a slowdown in replication can become critical. If it becomes significant, the follower partitions on the broker may fall out of sync with their corresponding leaders, leading to a reduction in the number of in-sync replicas for those partitions.

# Checking for an Imbalance of Partition Leaders

When checking for an imbalance of partition leaders among the brokers, it's sometimes not enough just to compare the total number of partition leaders between all the brokers. You should dive deeper and check per topic whether the partition leaders are balanced among all brokers. There are several reasons for that:

- There can be a partition leader skew while the brokers host the same number of partition leaders. Imagine a case when two big topics aren't balanced among the brokers, but the first topic has more partition leaders on one broker while the other topic has more leaders on another broker. In total, the number of partition leaders per broker would be the same, while if you look per topic you would see an imbalance.

- There can be a small difference in the number of partition leaders per broker, but this difference originates from a topic that has lots of traffic or many consumers.

The following example shows how even a slight skew can cause one broker to reach CPU saturation and cripple the Kafka cluster. Figure 3-2 shows a cluster with three brokers, one broker being very loaded (as can be seen from its load average).

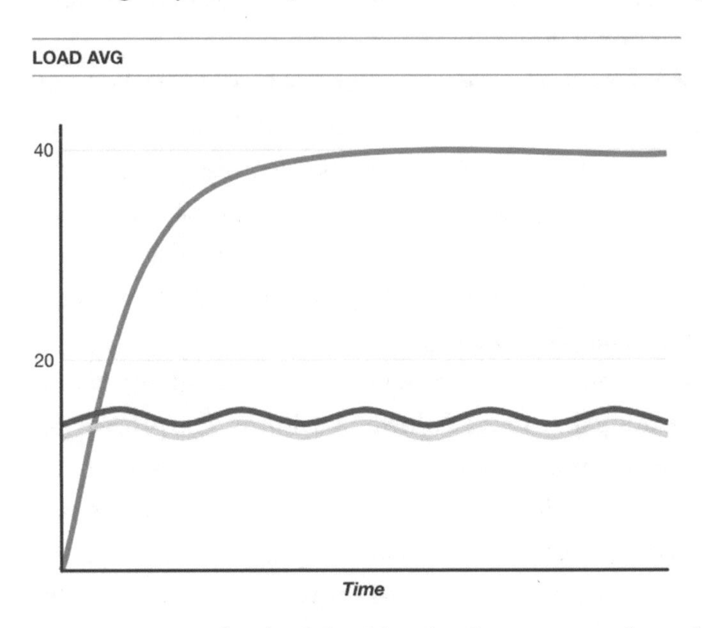


***Figure 3-2.*** *The load average (LA) of the blue broker is more than double than the LA of the other two brokers*

The cause for the high load average is the high CPU user time (called `us%`) of this broker, as shown in Figure 3-3.

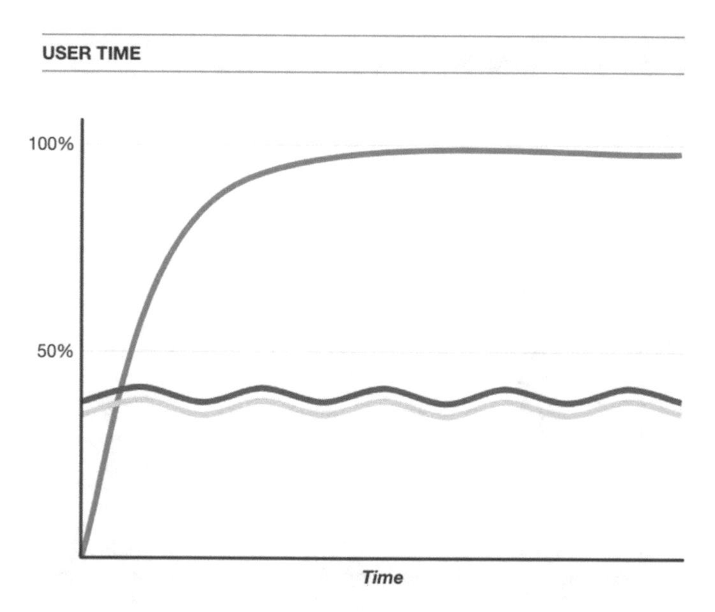


***Figure 3-3.*** *The CPU us% of the blue broker is more than double the us% of the other two brokers*

In searching for the cause of the high CPU us% in this broker, you can first check whether this broker receives or sends more traffic. However, you realize that all brokers had the same incoming and outgoing traffic.

So you would then compare the number of partition leaders per broker. You determine that there is very little variance in the number of leaders between the brokers, as shown in Figure 3-4.

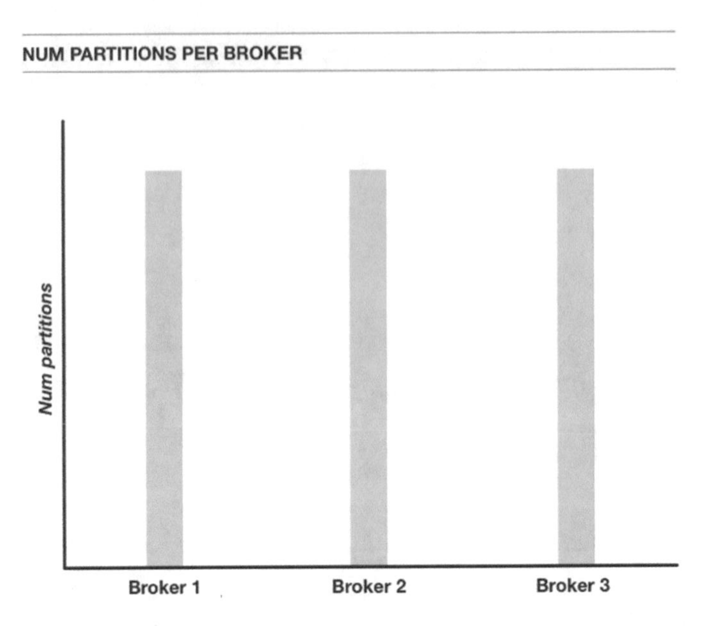


***Figure 3-4.*** *The number of partitions in all brokers is the same*

When you look deeper—at the scope per topic and not just per broker—you see a topic with a single partition leader that resides on broker number 2, as shown in Figure 3-5.

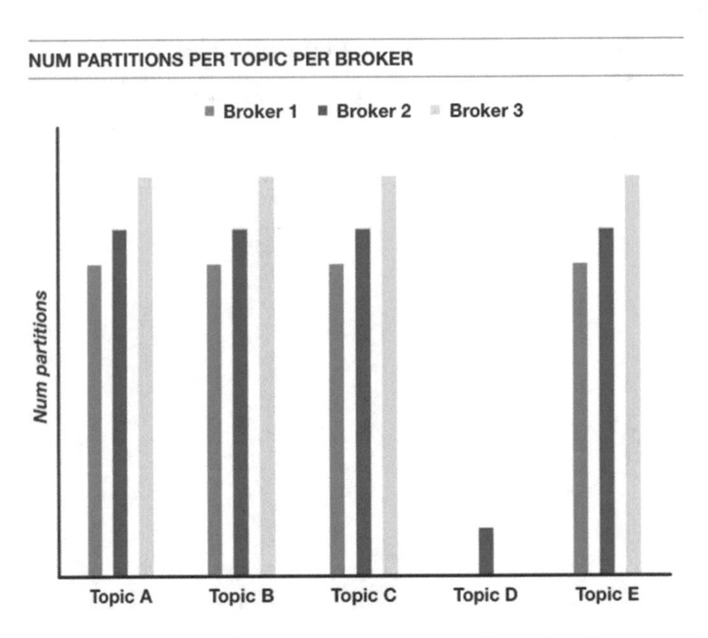


***Figure 3-5.*** *The number of partitions per topic and per broker. Topic D has a single partition that resides on broker 2*

At first glance, there's nothing special here—just a single partition leader that resides on some broker. However, that single partition leader was the only difference between the loaded broker and the other two brokers.

After further analysis, you find that the topic (to which this partition leader belongs) has much more consumers compared to the other topics on this cluster, as shown in Figure 3-6.

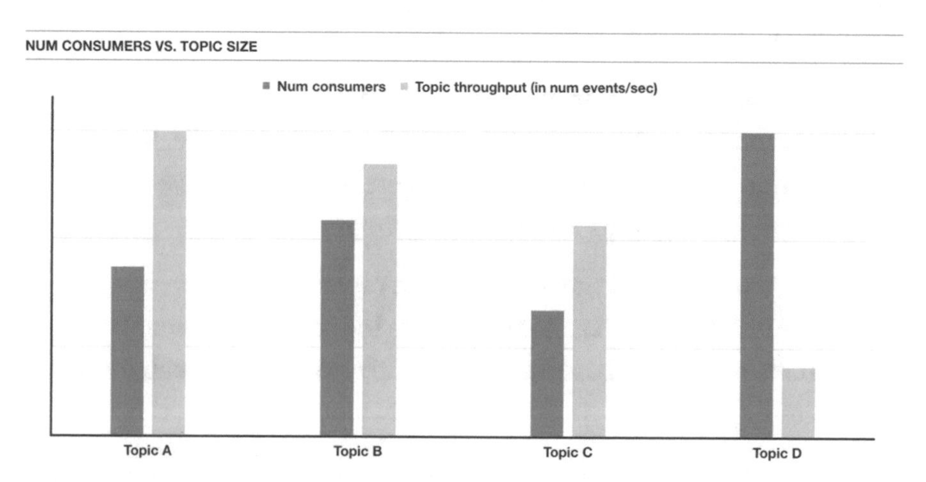


***Figure 3-6.*** *The number of consumers per topic and the throughput per topic*

The moral of this production incident (in the scope of this particular chapter) isn't that many consumers can cause high CPU and load average. Instead, the takeaway from this story is that when you're looking for partition leaders skew, it's not enough to just compare the number of partition leaders between all the brokers. You also need to check per topic whether their leaders are aligned among all the brokers. Otherwise, you can miss a case like the one illustrated here.

## Reassigning Partitions to Achieve an Even Distribution

In order to distribute partitions (of leaders and followers) evenly across the brokers, you need to reassign the partitions among the brokers. There are some aspects you need to consider when performing such a reassignment procedure; otherwise, it can degrade the performance of the cluster:

- *Network bandwidth:* During the reassignment, data can move between brokers, which might increase the network traffic. If the network is already a performance bottleneck, this can cause a delay in the network and thus become a performance issue.
- *Storage issue:* If data is transferred into a broker and the storage use (in the mount points that store the Kafka log segments) becomes full (100% of the storage is used), then the broker will probably stop functioning.
- *Disk I/O operations:* When data is being transferred from one broker to another, there are spikes in both read and write operations. For example, if partition P1 moves from broker B1 to broker B2, then the segments of P1 are being read from the disks of B1 (resulting in increased read operations on B1 disks) and written to the disks of broker B2 (resulting in increased write operations on B2 disks).
- *Consumer lags:* The reassignment causes not only an increase in disk operations, but also fills the page cache with data that's being transferred. For example, if partition P1 moves from broker B1 to broker B2, the segments of P1 are being read from the disks of B1 and then loaded into the page cache of B1. Once the partitions arrive to broker B2, they're written into the page cache of B2 and then flushed to the disks of B2. Usually, when the only data transfer between the brokers in the cluster is due to replication (there's no data transfer caused by partition reassignment), the consumers read

at least part of the data from the page cache. The impact of this on the consumers of the Kafka cluster is that the real-time data that flows into both brokers read data that doesn't exist in the page cache anymore (because it was flushed out by data that's related to the reassignment).

- *Transfer costs*: If the partitions are moved from brokers that reside on different AZs, there are costs associated with the data transfer.

# Data Distribution Among Disks

When a broker has more than one disk, you can end up having an uneven distribution of the data among these disks.

The reason is that, for partitions of new topics, the brokers will place these partitions on the disks that host the least number of partitions, instead of on the disks that have the least amount of used disk space.

In a perfect scenario, at any point in time, per each broker, the storage usage per disk will be the same. However, if for some reason the partitions aren't distributed evenly across the disks per broker (e.g., due to some partition reassignment that hasn't gone well), then some disks will end up having more heavy partitions then other disks.

The next time a new topic is created, more partitions will be stored on disks with a fewer number of partitions, and not with less storage usage. This can make the already uneven distribution of data per disk become even more uneven. In the worst case, this can lead to at least one disk in the Kafka cluster being full.

In order to prevent this case, whenever brokers have more than a single disk and there's a difference between the storage usage among the disks, check the distribution of partitions per topic among the disks. When you find a topic that has more partitions on those disks that have more used storage, perform partition reassignment for this topic.

# Summary

This chapter explored the complex problem of partition skew in a Kafka cluster, shedding light on the subtleties of both leader and follower skews and their effects on the system's balance and efficiency.

The chapter started by explaining what partition skew means, specifically focusing on leader and follower skews and the consequences of sending and receiving messages. When brokers are unevenly loaded, it can result in issues such as slow replication, consumption delays, and an increase in page cache misses.

The chapter then detailed the distinctions between leader and follower skews. Leader skew primarily affects the writing and reading of messages by producers and consumers, whereas follower skew has a significant influence on replication processes.

Next, the chapter looked into the problems that can occur when a broker manages too many partition leaders, such as higher disk I/O activities, delays in communication for both consumers and producers, an imbalance in followers, and a decrease in the number of synchronized copies.

We stressed the need to analyze skew not just on a broker-by-broker basis but also for individual topics. A case study was presented to demonstrate how just one partition leader skew can cause a CPU overload and substantial issues in the cluster.

The chapter also provided guidance on how to monitor and tackle skew challenges, taking into consideration aspects like the rate of messages, the number of consumers and producers, the status of in-sync replicas, and more.

Finally, the chapter delved into the particular concern of unequal data distribution among a broker's disks, emphasizing the possibility of storage imbalances and outlining the measures that can be implemented to alleviate these difficulties.

The next chapter delves into the intricate issue of skewed or lost leaders in Kafka, an area that has significant implications for brokers, consumers, and producers alike. As you explore the mechanisms of Kafka, understanding how these leaders can become skewed or lost and the consequences of such irregularities becomes paramount. You'll unravel the underlying factors that may lead to these problems, such as networking issues or challenges with partition leadership. Additionally, you'll learn practical solutions and preventive strategies to ensure that the Kafka system operates smoothly and efficiently.

# CHAPTER 4

# Dealing with Skewed and Lost Leaders

This chapter explains how to detect and troubleshoot issues related to skewed or lost leaders. In Kafka, messages are produced into topics by producers. These messages are later consumed from these topics by consumers. Figure 4-1 shows a general sketch of this flow.

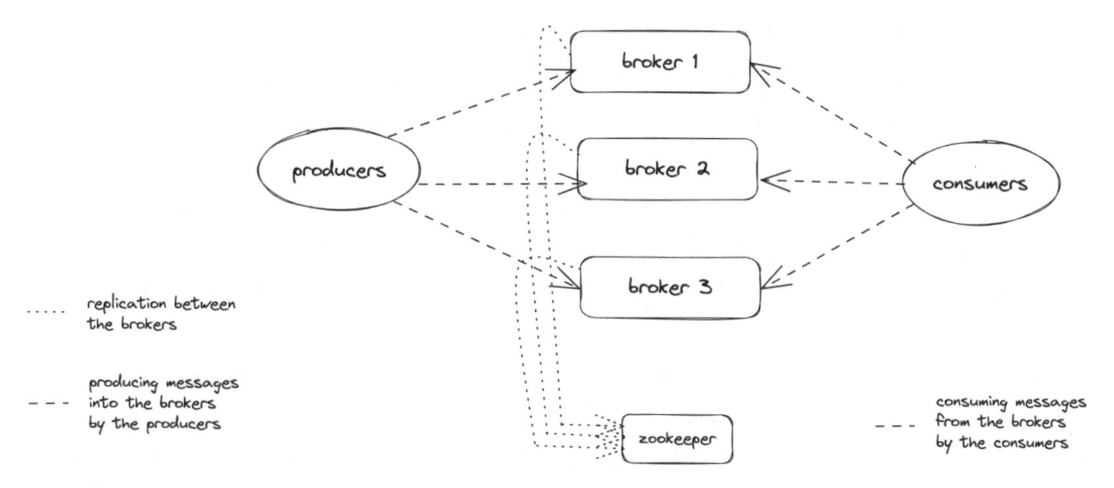


***Figure 4-1.*** *The general flow of messages from producers into brokers and then to consumers*

A topic is broken into multiple partitions, so you can think of the relationship between a topic and its partitions just like folders in a filesystem—a topic is a folder and inside this folder there are more folders, each a separate partition. Each partition has a leader broker. This leader broker is in charge of all the read and write operations for that partition. Also, each partition leader has one or more followers, depending on the replication factor of the topic that the partition belongs to.




E. Eldor, *Kafka Troubleshooting in Production*, https://doi.org/10.1007/978-1-4842-9490-1_4

A leader skew has different meanings, depending on which scope you look at. It indicates a situation where certain brokers are leading a greater number of partitions compared to their counterparts. However, if you focus on a specific topic, a leader skew means that, for that particular topic, some brokers are leading more partitions than others. It's important to note that in this scenario, the skew only pertains to the partitions related to that specific topic.

A partition leader can also get lost if the broker that serves as its leader stops serving as its leader for some reason. This becomes a stability issue for the cluster if many leaders lose their leadership at once, or when this occurs too frequently.

These two symptoms—leader skew and lost leaders—can adversely impact the brokers, consumers, and producers.

# When Partitions Lose Their Leadership

While working with Kafka consumers, there were multiple times in which I saw these consumers suffer from backlogs, crashes, or just hang. When correlating these symptoms with the Kafka logs, I noted that, just before these symptoms occurred, several leaders were lost.

Let's first define a lost leader. Each partition in a topic has exactly one partition leader, which handles all the read/write requests of that partition. If that leader partition is no longer the leader, then another partition (out of the follower partitions of this leader) becomes the leader.

In this section we won't deal with the issue of some partition leaders losing their leadership sporadically. Instead we'll tackle the case in which many leader partitions from different topics lose their leadership at once, all in the span of a few seconds.

Why does this happen? At first, I tended to suspect a faulty broker; however that wasn't the case, since the partitions that lost their leadership always resided on several brokers.

After a while, I found that the massive load of partitions losing their leaders was caused by one of two networking issues—related to the ZooKeeper or the NIC.

Before delving into each of these issues, let's first elaborate on both components.

## ZooKeeper

The ZooKeeper (aka ZK) appoints a leader to every partition, and it is in charge of managing all associated read and write requests. It also nominates a controller broker with the responsibility of overseeing changes in the state of brokers and partitions. Furthermore, It's tasked with maintaining a continuously updated record of metadata for the Kafka cluster, encapsulating data on topics, partitions, and replicas.

This first networking issue relates to the session timeout of the ZooKeeper. There are several steps to determine and mitigate this problem.

- First, you need to check whether the ZooKeeper is suffering from high garbage collection (GC). If it is, you can address this by increasing the amount of RAM allocated to the ZooKeeper until these GC issues are resolved.
- If GC is not an issue, you can proceed to the next step, which involves examining the ZooKeeper's session and connection timeouts. These timeouts may be too short, and if so, you can try to increase the session timeout.

Once you've made these changes, restart ZooKeeper and see if the leader partitions still lose their leadership. In my experience, once you increase the ZooKeeper's session timeout, the frequency of leader partitions losing their leadership diminishes significantly. Consequently, the occasions when consumers hang due to lost leaders also decreases substantially.

## The Network Interface Card (NIC)

A Network Interface Card (NIC) is a hardware component that facilitates the connectivity of servers (and in this case, Kafka servers) with the network and enables the Kafka server to send and receive data over the network. The second networking issue pertains to NIC saturation.

This issue is something I've encountered exclusively on on-premises clusters, not on cloud environments. Although it's less common than the ZooKeeper session timeout issue, it still happens occasionally. When the traffic causes saturation of the NIC on the Kafka brokers, the NIC can reset. So, when you witness a loss of leaders, it's worthwhile to check the `dmesg` logs for any NIC-related errors. If the loss of leaders happens exclusively on brokers in which their NIC was reset under load, you should examine your NIC configuration. In such scenarios, simply increasing the ZooKeeper session timeout wouldn't be beneficial.

If your Kafka consumer application is either hanging, accumulating a backlog, or crashing when reading from a Kafka topic, or your Kafka producers are unable to produce into a Kafka topic, you might be experiencing issues with session timeouts or some other causes.

One of the common causes is that Kafka brokers cannot reach the ZooKeeper due to session timeouts. Alternatively, the issue might stem from the ZooKeeper dealing with long or frequent full garbage collection pauses. Occasionally, the NIC in some Kafka brokers might also reset when under a high load. There are other potential reasons, such as broker restarts/crashes and network partitioning, but we won't get into their details in this section.

To prevent such scenarios, ensure that the ZooKeeper session timeout isn't too short. Additionally, take steps to ensure that the NIC in the Kafka brokers does not reset under high load.

For optimal system health, you should continuously monitor full GC activity in the ZooKeeper, look out for resets of the NIC, and keep an eye out for errors related to lost leaders.

# Should Leader Skew Always Be Solved?

A Kafka cluster is said to have leader skew when there's at least one broker with more partition leaders than the average number of partition leaders in the cluster (number of leaders divided by number of brokers). The popular recommendation in such a case is to reassign the leaders of all the skewed topics so that each broker has the same number of leaders.

However, there are several things to consider when there's leader skew.

A cluster can suffer from leader skew even if all the brokers have the same number of leaders, since even in such a case there can be one topic with leader skew that's offset by leader skew in the other direction. That's why it's important to check for leader skew not just in the scope of the whole cluster, but also in the scope of specific topics. In order to handle a leaders' skew in such a case (when there's no partition skew), you can run a manual leader election.

In order to fix partition skew, you need to run a partition reassignment procedure, which not only takes time but also increases the CPU utilization of the brokers (due to the higher data transfer that occurs during the reassignment), especially with CPU user time (`us%`) and disk I/O time (`io%`). This can sometimes be problematic for Kafka clusters that already suffer from high CPU, so there are times when we'll want to avoid partition reassignment even if there's a leader skew. Alternatively, we can drop the retention on the topics that contribute to most of the traffic in the cluster in order to run the partition reassignment process more smoothly.

That's why it's important to emphasize that not all leaders' skew should be dealt with. However, there are times when it's crucial to solve the leaders' skew, and these are the cases in which a leaders' skew in some topic has a real effect on the Kafka clusters. There are two such issues—high traffic and a high number of consumers/producers from this topic. The following sections consider these two cases.

## When There Is High Traffic

When a topic has high incoming throughput, a leaders' skew can cause the brokers that host more leaders of this topic to suffer from a higher CPU usage, which can subsequently cause high network threads usage in these brokers. The high CPU and network threads usage can in turn cause consumer backlog of stalled producers.

Consumers can backlog because when a Kafka broker experiences high CPU usage, it becomes less efficient at processing the requests it receives. This means it takes longer for the broker to respond to consumers' fetch requests, which are the requests consumers make to retrieve messages from the broker.

On the consumer side, it expects to fetch messages at a certain rate. If the broker is slow to respond due to high CPU usage, the consumer won't receive messages as quickly as it expects. As more messages are produced to the topic, and the broker continues to be slow to respond, the consumer will start to lag behind the production of new messages. This is what is meant by a consumer backlog—the consumer isn't keeping up with the incoming data rate, so there's a growing backlog of messages that the consumer hasn't fetched and processed yet.

Similarly, for producers, if the broker is slow to acknowledge the receipt of new messages due to high CPU usage, the producers may be stalled waiting for the acknowledgements, which can slow the overall message production rate.

## When There Is a Large Number of Consumers/Producers

Even a topic with small throughput that has many producers or consumers can cause the brokers that host more of its leaders to suffer from high CPU.

When a Kafka cluster has a leader skew, the recommendation is to perform a partition reassignment (or a manual leader election), but only for those topics that either have high traffic or high number of consumers and/or producers. In other cases, check whether the brokers that host most of the leaders suffer from either higher CPU usage or have more storage. If they don't, you can leave the cluster as is without performing leader reassignment. Figure 4-2 shows an example of the distribution of partition leaders and followers for three partitions of a specific topic in a Kafka cluster.

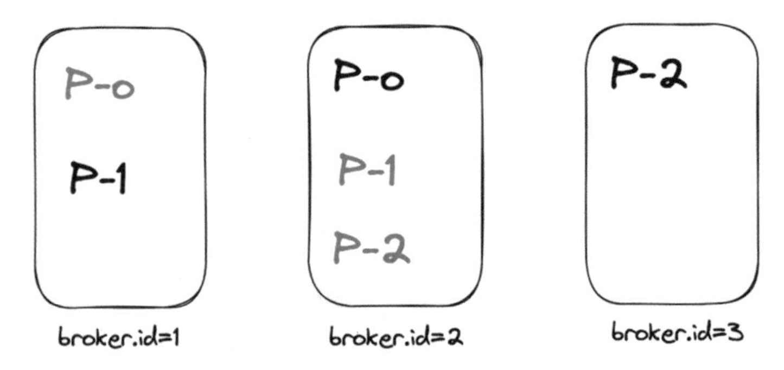


P-i - follower partition, e.g. - the follower of partition number 1 is the broker with broker.id=1

P-i - leader partition, e.g. - the leader of partition number 1 is the broker with broker.id=2

***Figure 4-2.*** *An example of the distribution of partition leaders and followers (P-0, P-1, P-2) between three brokers*

## Understanding Leader Skew

The implications of leader skew were discussed in the previous section; however this section takes some time to better define leader skew.

Formally, *leader skew* is defined by a non-even distribution of leaders among the brokers, as shown in Figure 4-3.

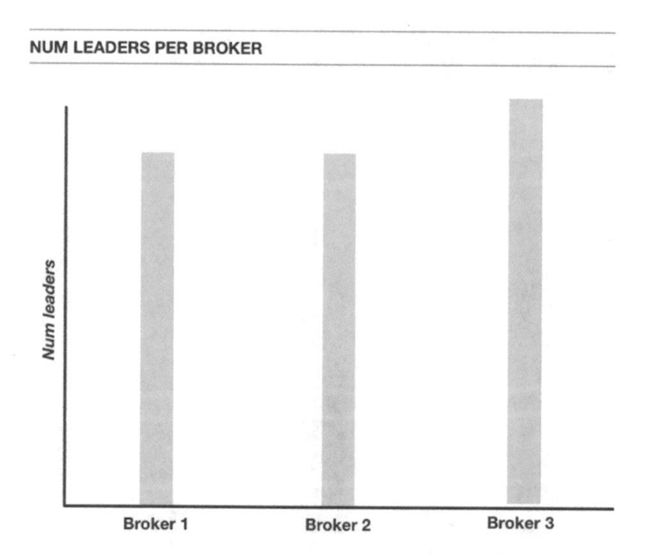


***Figure 4-3.*** *A skew of partition leaders among the brokers*

There's a pitfall to using this approach to identify leader skew, since it's not per topic but instead per broker.

Knowing the leader skew isn't enough; you need to understand which topics contribute to this skew. When looking at the number of leader partitions per broker *and per topic,* as shown in Figure 4-4, you can see the following anomaly—Topic D has a single leader partition that resides on Broker 2. This was a real case in which topic D caused Broker 2 to reach high CPU user time due to many consumers reading from the leader partition of that topic.

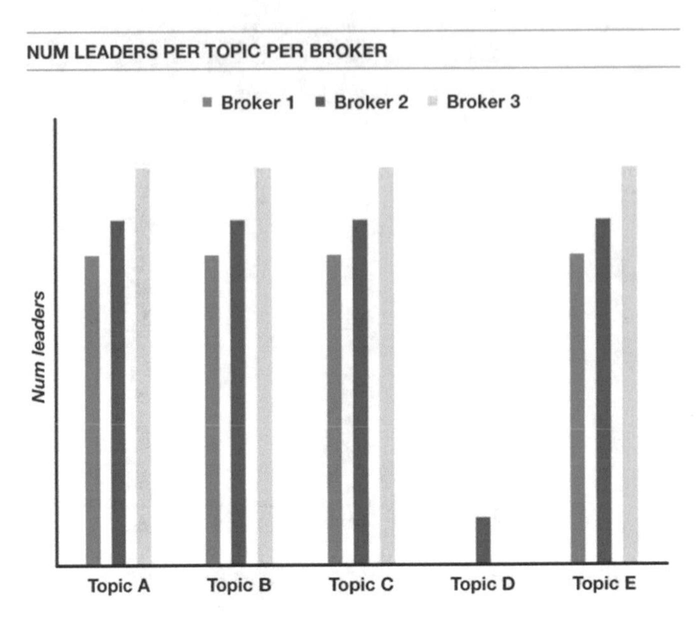


***Figure 4-4.** The distribution of leader partitions per broker and per topic*

Figure 4-5 shows another example of uneven distribution of leaders per brokers.

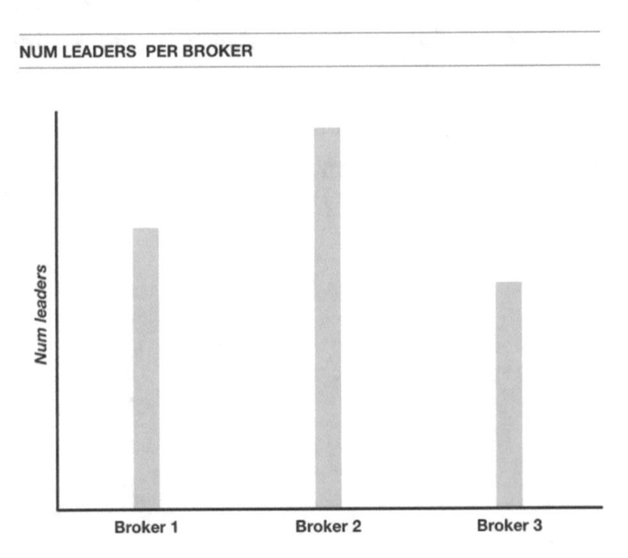


***Figure 4-5.** An uneven distribution of leader partitions among brokers*

This skew of leader partitions doesn't tell you much until you dive deeper into which topics contribute to the skewed distribution, as shown in Figure 4-6.

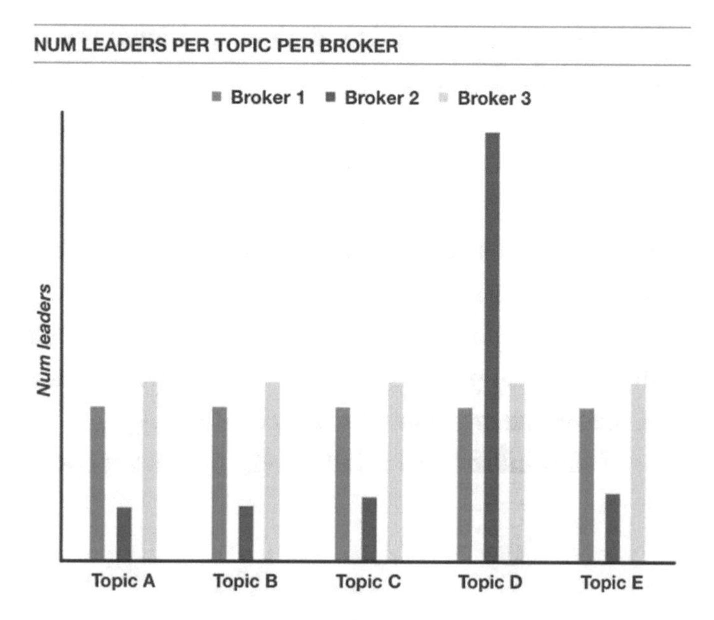


***Figure 4-6.*** *The number of leader partitions per topic and per broker*

In this case, you can see that most of the skew can be attributed to Topic D.

# Summary

This chapter explored the impact of skewed or lost leaders in Kafka on brokers, consumers, and producers. In Kafka, each partition has a leader broker responsible for read and write tasks, and these leaders can sometimes become skewed or lost. This skew or loss, depending on whether it's across all topics or a specific one, can cause problems for both consumers and producers, like instability, crashes, and backlogs.

The root cause of these issues sometimes lies in networking problems related to ZooKeeper, which is the entity assigning partition leaders, or the Network Interface Card (NIC), which connects servers to the network. Potential solutions include increasing the RAM allocated for the ZooKeeper in order to mitigate high garbage collection (GC) or extending its session timeouts, and checking for NIC-related errors leading to necessary configuration adjustments.

Leader skew is a situation where one broker leads more partitions than the average. Although reassigning leaders for equal distribution is a common solution, it might not always be the best approach due to the increased CPU usage and time consumption.

Nonetheless, addressing skew is crucial for topics with high traffic or many consumers/producers. High traffic and a large number of consumers/producers can lead to increased CPU usage in brokers hosting more leaders, causing issues such as consumer backlog and stalled producers. Therefore, partition reassignment or manual leader election is recommended in such cases.

The next chapter dives into the topic of CPU saturation in Kafka clusters. We'll learn what CPU saturation is and how it differs from a fully engaged CPU, cover the various types of CPU usage, and explore how each one can affect Kafka. The chapter also discusses the influence of log compaction and the number of consumers per topic on CPU usage, including real-world examples and practical strategies. The goal is to provide a clear understanding of CPU utilization in Kafka, helping you spot potential issues and apply effective solutions.

CHAPTER 5

# CPU Saturation in Kafka: Causes, Consequences, and Solutions

There are cases in which CPU saturation can cause the Kafka cluster to delay consumers and/or producers, by either causing consumers to lag or causing producers to develop a bigger queue of messages before producing them to Kafka.

In such cases, it's important to identify which kind of CPU saturation you're dealing with, because this can provide clues to its cause.

We'll start by describing what CPU saturation is and explain the different CPU usage types. We'll include several examples, each causing a different type of CPU saturation.

## CPU Saturation

A Kafka cluster includes several machines, each of which runs a single Kafka process. When the Kafka process generates more tasks (tasks that require CPU) than the number of available CPU cores, these tasks will wait in the OS queue for their turn until the CPU cores become available. This waiting can cause latency for the brokers and its consumers and producers, leading to a performance degradation of the Kafka cluster.

The demand for CPUs is indicated by an OS metric called *load average,* which refers to the number of tasks that are either running or waiting to be run by the CPUs (once a thread is scheduled to run on the OS, its executable code is considered to be a task).

When the load average is higher than the number of CPUs, it means the CPUs are saturated. The runnable threads must wait for the CPUs to become available.




E. Eldor, *Kafka Troubleshooting in Production*, https://doi.org/10.1007/978-1-4842-9490-1_5

The term *normalized load average* (NLA) describes the levels of saturation. For example, in a 16-core machine that has a load average of 16, the NLA is 1 (because 16/16=1). In such a case, all runnable tasks are being served, and no runnable task is waiting in the queue. However, when the load average is 32, the NLA is 2 (since 32/16=2). In such a case, there are 16 tasks running on the CPUs; the other 16 tasks are in a runnable state and are waiting for a CPU to become available.

It's important to distinguish between a fully engaged CPU and CPU saturation. CPU saturation is when all the processing units (cores) in a CPU are completely engaged and operating at maximum capacity.

When CPU utilization reaches 100 percent, this means that every core is actively processing tasks, and there is no idle time left for any new tasks to be assigned to the cores. In this situation, the CPU is fully engaged.

However, even when a CPU is fully engaged, that doesn't necessarily imply CPU saturation. The CPU can still accommodate more tasks without causing any observable performance degradation if those tasks are not demanding or if they have a low priority. In other words, a fully engaged CPU is merely a CPU that is efficiently using all of its resources, not necessarily one that is overwhelmed or overworked.

When the Normalized Load Average (NLA) exceeds 1, this indicates CPU saturation as it represents the queueing of processes. The load average is a measure of the number of runnable tasks, including the tasks that are running and the ones that are waiting to run (i.e., queued). When the NLA is greater than 1, this implies that there are more runnable tasks than available cores. This waiting or queueing of tasks signifies CPU saturation, as the CPU doesn't have enough resources to immediately accommodate all tasks. This means the tasks are waiting, which can potentially lead to slower system response and processing times.

The key difference is that a fully engaged CPU has all its cores actively working and can still process incoming tasks without delay, while CPU saturation occurs when there are more tasks to be processed than the CPU can handle at once, leading to queuing and potentially slowing down task processing.

I've noticed that when Kafka consumers and/or producers experience latency due to CPU saturation, it's usually not because the CPUs have reached 100 percent utilization. Usually, it's due to the NLA reaching a value higher than 1.

Given that, the more interesting question becomes this—what causes tasks to queue (which is equivalent to asking why the NLA is higher than 1) when CPU utilization is below 100 percent?

To answer this question, let's first overview the different CPU usage types, since detecting which type of CPU reaches a high utilization can assist in troubleshooting latency issues.

## CPU Usage Types

Linux CPU usage can be broken into several distinct categories:

- *User time,* represented as `%us`, measures the portion of CPU time engaged in executing processes in the user space, usually tied to application-related tasks.
- *System time,* denoted as `%sy`, indicates the part of CPU time spent in the kernel space, working on tasks vital to the system's functioning.
- *I/O wait time,* or `%wa`, signifies the amount of CPU time spent waiting for input/output tasks, typically associated with disk operations.
- S*oftware interrupts,* labeled as `%si`, highlights the segment of CPU time dedicated to managing these types of system disturbances.

Now that we have an idea what CPU saturation is and know the different categories of CPU usage in Linux, the next section explains what causes Kafka brokers to use CPU time.

## Causes of High CPU User Times

This section covers the possible causes of high CPU user time, along with ways to monitor and potentially prevent these cases.

Exploring the factors leading to elevated CPU usage provides insights into how to monitor and mitigate such situations effectively. High CPU usage can result from numerous fetch requests coming from consumers or lots of produce requests from producers. Additionally, a high volume of fetch requests from other brokers, a part of the replication process, can also contribute to high CPU usage.

The brokers' compression settings can further impact CPU usage. If brokers are set to compress or decompress messages, particularly when both producers and brokers utilize compression, this can increase CPU usage. Message encryption and decryption by brokers can also affect CPU usage.

Garbage collection (GC) is another potential source of CPU stress. Frequent or lengthy GC pauses can significantly increase CPU usage.

To prevent high CPU usage, we can employ several strategies. We can experiment with different compression methods and observe their impact on the CPU. If brokers are configured for encryption, it is worth assessing whether this step is essential.

It is also advisable to verify whether the producers and/or consumers are already configured to compress or decompress. If so, there is no need for the Kafka brokers to perform the same function. Regarding GC, keep an eye on GC pauses and consider increasing the broker's heap size if needed. Using the G1 garbage collector (enabled with `-XX:+UseG1GC`) can also be beneficial.

Monitoring should encompass the number of consumers and producers per topic, along with the GC log (enabled with the `-XX:+PrintGC` flag for the broker's process). Look for frequent or long GC events. Moreover, it's a good idea to set up an alert for instances when CPU usage exceeds 80 percent for durations longer than ten minutes.

## Causes of High CPU System Times

This section covers the possible causes of high CPU system times, along with ways to monitor and potentially prevent these cases.

High CPU system time, or `sy%`, can result due to several factors. For instance, when `linger.ms` and/or `batch.size` are set to low values, the number of produce requests coming into the cluster might increase. This is because smaller linger and batch values prompt the producer to send messages to the brokers more quickly, reducing the wait time for these messages in the producer's buffer. Consequently, a surge in produce requests can lead to a hike in CPU system use.

Excessive access to disks and a high number of configured I/O threads in the cluster can also push the CPU `sy%` up. A general rule of thumb is to configure the I/O threads up to the number of mount points per broker used to store segments. In addition, hosting a compacted topic with high retention in the cluster can also escalate CPU use.

To prevent a high CPU `sy%`, examine the `batch.size` and `linger.ms` settings in the producers since they might be too low. Also, evaluate whether the number of disk I/O threads is overly high. Further, review your topics. If any are compacted, ensure their retention is not higher than necessary.

For monitoring purposes, set an alert for situations when the CPU sy% exceeds 20 percent for more than ten minutes. This can serve as a preventative measure against prolonged system stress, helping to maintain optimal system performance.

## Example of Kafka Brokers with High CPU %us and %sy

Here is an example of a spike in the CPU us% (user time) and sy% (system time) in all the brokers during peak times. The consumers suffered from lag due to the high CPU sy%. Note that the lag can also be caused by high CPU percent. Figures 5-1 and 5-2 illustrate this example.

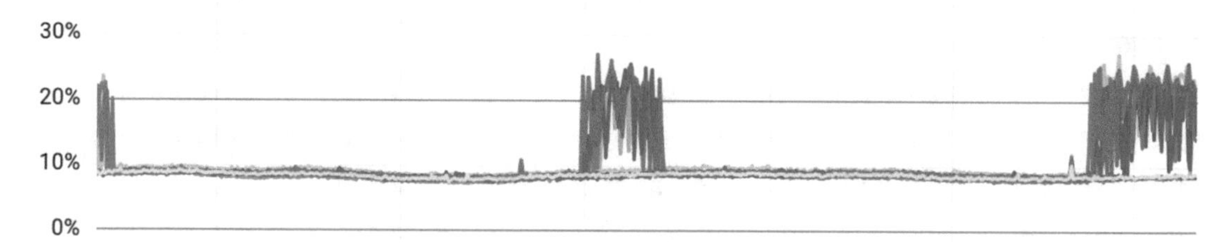


***Figure 5-1.** A spike in the CPU sy% during peak time*

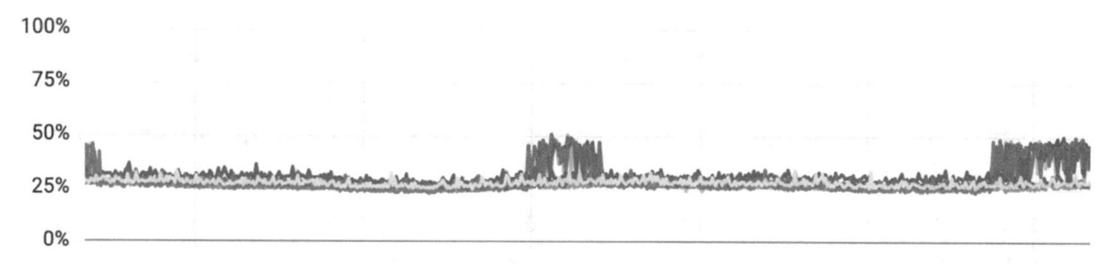


***Figure 5-2.** A spike in the CPU us% during peak time*

An example of a constantly high CPU sy% (of about 20 percent) is shown in Figure 5-3, in all the brokers of the Kafka cluster. The consumers of this cluster sometimes suffer from lag, but only during peak times. In this case, this high CPU sy% was caused by disk contention, which subsequently caused consumers to lag during peak time.

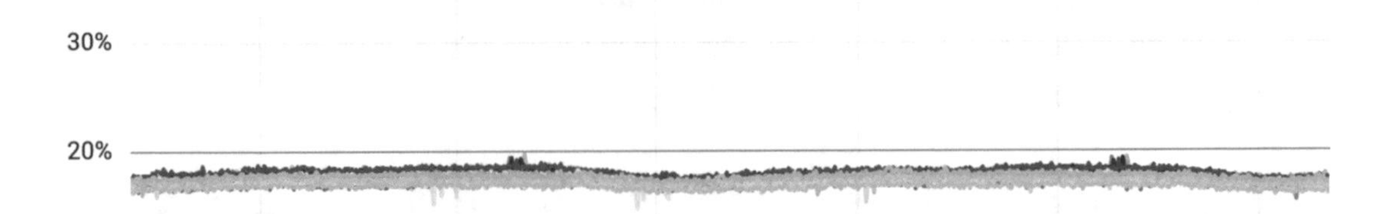


***Figure 5-3.*** *A constant high value of 20 percent in the CPU system time*

## Causes of High CPU Wait Times

This section covers the possible causes of high CPU wait times, along with ways to monitor and potentially prevent these cases.

Several factors can lead to a high CPU wait time. For example, a sudden surge in read or write operations from the disk, surpassing the disk's capacity in input/output operations per second (IOPS), can cause an increase in CPU wait time (`wa%`). Other potential causes include consumer lag, a malfunctioning disk, or a disk that cannot handle the required IOPS or throughput.

If you notice high CPU `wa%`, you should investigate two other metrics—disk utilization percentage and page cache miss ratio. A high value in either of these metrics could be the culprit behind the increased CPU `wa%`. If disk utilization is high, determine whether it's caused by throughput or IOPS. On the other hand, if the page cache miss ratio is high and there are no lagging consumers, it may suggest that the broker machine needs more RAM.

In terms of monitoring, establish an alert for scenarios when the CPU `wa%` exceeds 20 percent for more than ten minutes. This proactive measure can help you identify and mitigate potential system issues, maintaining the system's overall performance.

## Causes of High CPU System Interrupt Times

This section covers the possible causes of high CPU system interrupt times, along with ways to monitor and potentially prevent these cases.

A few factors can result in high CPU system interrupt time. For instance, an excessive number of disk.io threads, faulty or outdated device drivers, or hardware issues like defective disks or network cards can all lead to an increased number of software interrupts.

To prevent these issues, you should ensure that the number of disk.io threads is kept to a minimum, ideally no larger than the number of disks in the broker. A lower number of threads can help prevent system overload and unnecessary software interruptions.

For effective monitoring, consider setting an alert for scenarios where CPU software interrupt time (`si%`) surpasses 5 percent for more than ten minutes. This approach can help you promptly identify and address system performance issues, thereby enhancing overall efficiency.

# The Effect of Compacted Topics with High Retention on Disk and CPU Use

This section explains how a seemingly simple feature such as a topic's log compaction can bring your Kafka cluster to a halt while the compaction runs.

## What Is Log Compaction?

*Log compaction* is a process in Kafka whereby, for each key, only the most recent value is retained. It can be configured per topic and runs periodically in the background. This mechanism of storing only the most recent value for each key effectively removes duplicate records. Moreover, it also removes keys associated with null values, commonly referred to as *tombstone records.*

Notably, log compaction operates at the segment level. Once a segment crosses a certain threshold, known as the *dirty ratio,* Kafka flags it for cleaning. The dirty ratio refers to the proportion of records in a log segment that are obsolete and can be discarded.

In this cleaning process, records, including tombstone records, are copied to a new and clean segment. Tombstone records are special messages produced by consumers to delete keys, not for updates. They have the same key, but a null value. These tombstone messages have a separate retention policy that allows consumers to read the old records for a certain duration after the key deletion.

After the copying process, the new, clean segment is swapped in place of the old one, which is subsequently deleted. This entire log compaction process requires RAM and CPU cycles on the brokers, emphasizing its influence on resource utilization.

Through log compaction, Kafka ensures a more efficient storage utilization by maintaining only the most recent value for each key, while still enabling consumers to read old records during the tombstone retention period.

Figure 5-4 shows how log compaction works.

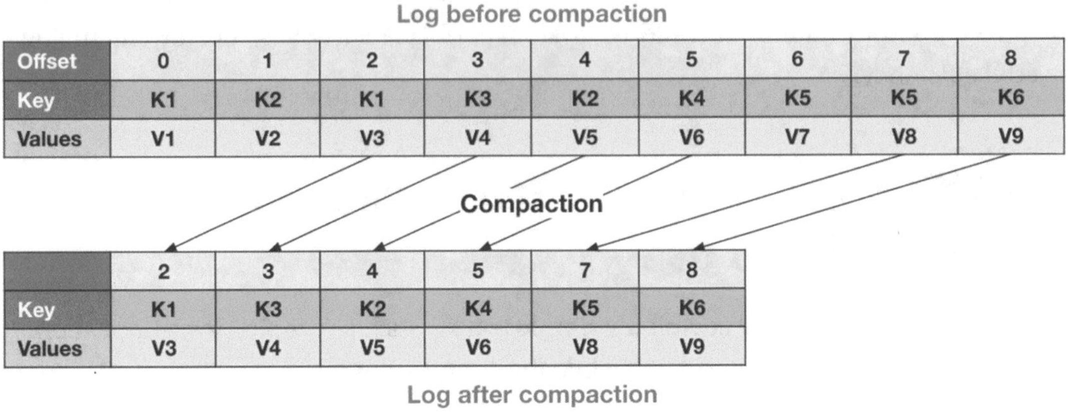

Log before compaction

| Offset | 0 | 1 | 2 | 3 | 4 | 5 | 6 | 7 | 8 |
|---|---|---|---|---|---|---|---|---|---|
| Key | K1 | K2 | K1 | K3 | K2 | K4 | K5 | K5 | K6 |
| Values | V1 | V2 | V3 | V4 | V5 | V6 | V7 | V8 | V9 |

Compaction

| | 2 | 3 | 4 | 5 | 7 | 8 |
|---|---|---|---|---|---|---|
| Key | K1 | K3 | K2 | K4 | K5 | K6 |
| Values | V3 | V4 | V5 | V6 | V8 | V9 |

Log after compaction

***Figure 5-4.*** *A topic with log compaction enabled. After the log compaction runs, only a single value per each key is saved*

Usually the log compaction process doesn't cause a CPU burden on the cluster, but a compacted topic with large enough retention might cause all brokers in a cluster to saturate due to a high CPU usage for part of the time. This can cause the cluster to function poorly during those times. The following section describes such a scenario.

## Real Production Issues Due to Log Compaction

Consider a cluster that suffered from the following symptoms.

The utilization of all the cluster's disks reached 100% and stayed there, as shown in Figure 5-5.

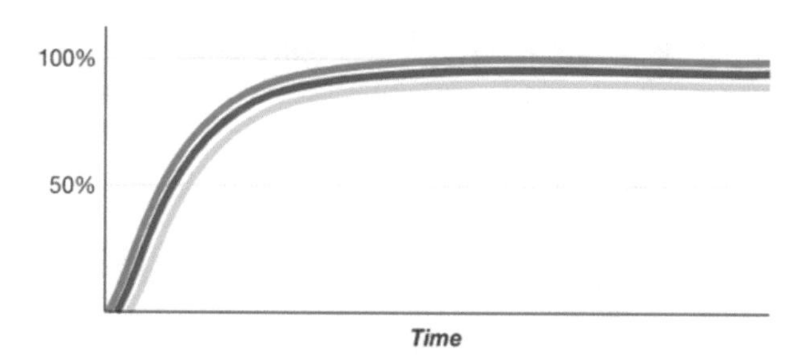


***Figure 5-5.*** *Disk utilization (a metric that was taken from the iostat tool) in all the brokers reached a value of 100% and stayed there. This means that all the disks in this cluster are saturated*

The CPU's system time (sy%) was ±15%, as shown in Figure 5-6.

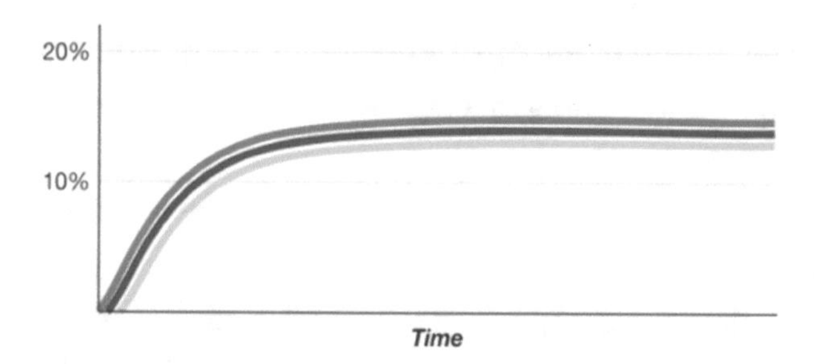


***Figure 5-6.*** *CPU sy% in all the brokers reached a value of 15% and stayed there. Such a value is high for Kafka clusters*

The Load Average was higher than the number of cores, which means there are more Kernel tasks that are either in status waiting or running, as shown in Figure 5-7.

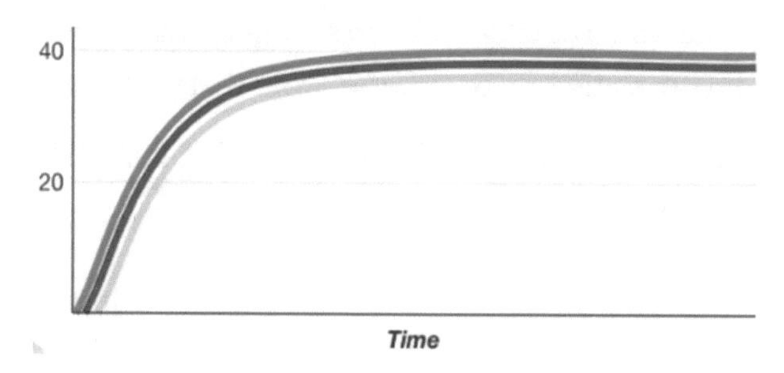


***Figure 5-7.*** *Load Average reached a value of 40 and stayed there. Since the number of cores in the brokers is 32, at any given time tasks wait in the queue for the CPU to run them. The Normalized Load Average of all the brokers is 1.25 (since 40/32 = 1.25)*

However, the CPU us% was low, as shown in Figure 5-8.

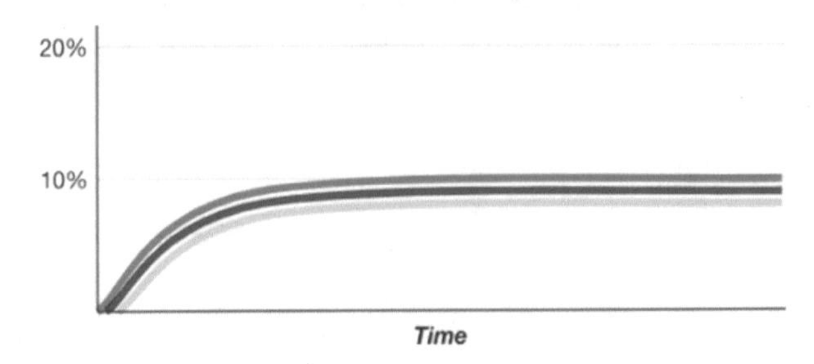


***Figure 5-8.*** *The CPU us% was low on all the brokers*

After several tries, the developer who created the cluster noticed that the retention of the compacted topic was 24 hours, which was much higher than required. Once the retention was reduced to 1 hour, the disk utility usage was drastically reduced, the CPU sy% was reduced by half, and the load average was reduced to less than the number of cores.

The conclusion is that when you use a large, compacted topic, pay attention to the retention. A high retention can be a possible cause for high disk utilization and high CPU sy% values during the time the compaction runs, which can cause consumer and/or producer lags due to the cluster struggling with the high load average.

# The Number of Consumers per Topic vs. CPU Use

I had a case in which consumers started lagging and messages started to queue in the producers. I looked at the Kafka cluster and saw a single broker that went rogue (out of a total of three brokers). The rogue broker suffered from these two symptoms:

- The Load Average was much higher compared to the other brokers, as shown in Figure 5-9.

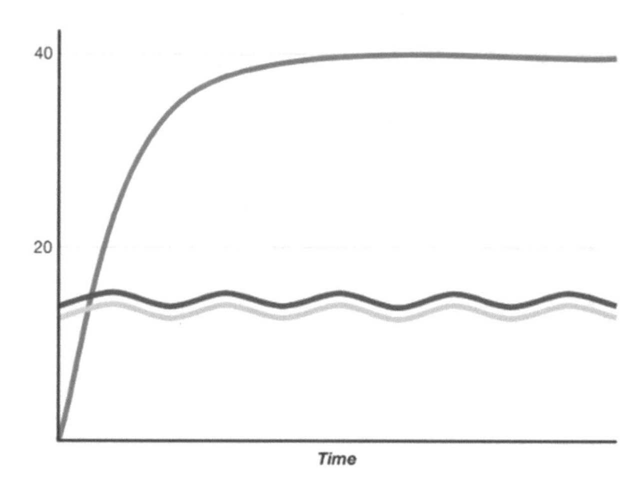


***Figure 5-9.*** *Load Average reached a value of 40 only in the rogue broker, which was more than double the load average of the other two brokers in the cluster*

- The CPU us% was much higher compared to the other brokers, as shown in Figure 5-10.

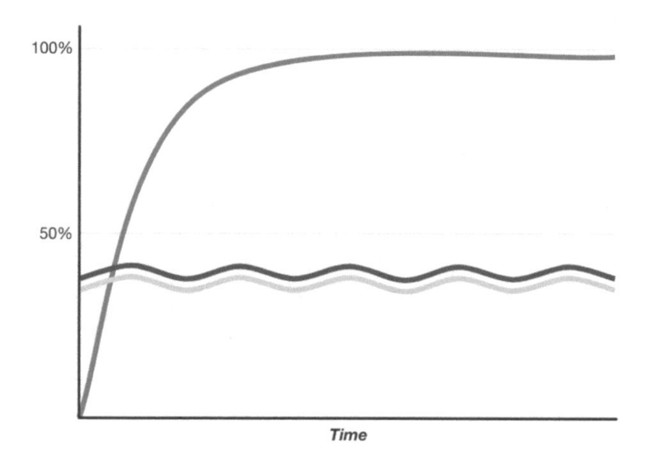


***Figure 5-10.*** *CPU us% reached 100% in the rogue broker, which was more than double the us% of the other two brokers in the cluster*

At first, I suspected that this broker either received and/or sent more traffic. However, it turns out the traffic was the same in all brokers, both in and out, as shown in Figures 5-11 and 5-12.

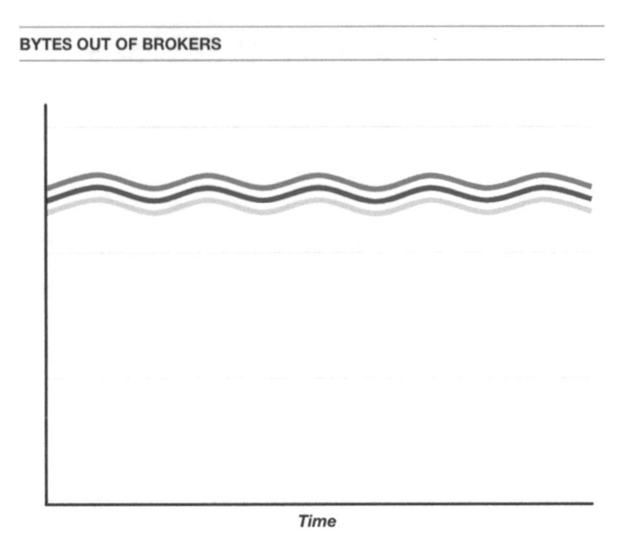


***Figure 5-11.*** *The number of bytes that were sent from all the brokers was almost the same*

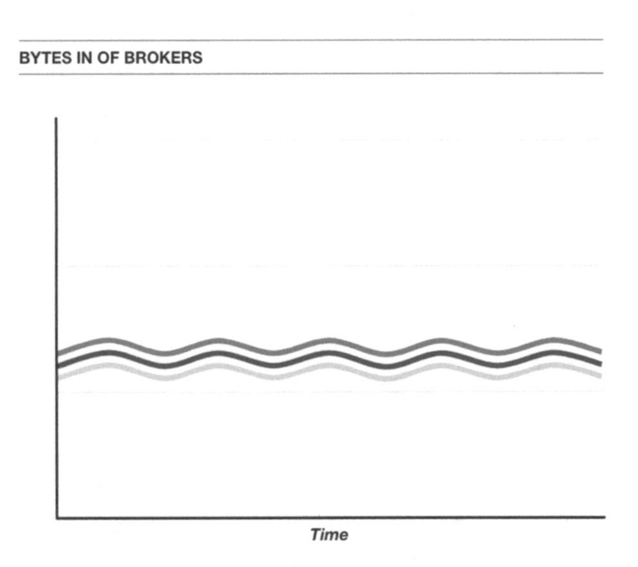


***Figure 5-12.*** *The number of bytes that were received by all the brokers was almost the same*

The next step was to suspect a partition skew—maybe this broker contained more partitions than the other brokers. However, it turned out that all the brokers had almost the same number of partitions, as shown in Figure 5-13.

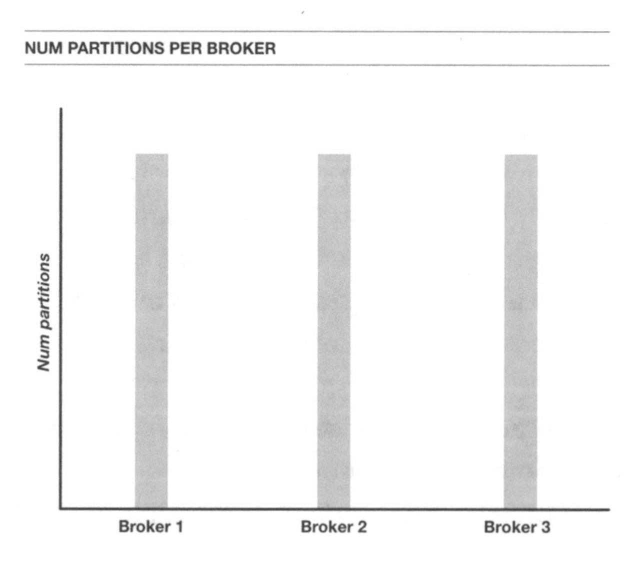


***Figure 5-13.*** *The number of partitions in all the brokers was almost the same*

At that point I was clueless, until a developer who was working on that cluster found the issue—there was a single topic, with low incoming traffic, that had many consumers.

What was more surprising is that the topic had only one partition, and that partition resided on... the rogue broker! Figure 5-14 shows that the single partition resided on the rogue brokers.

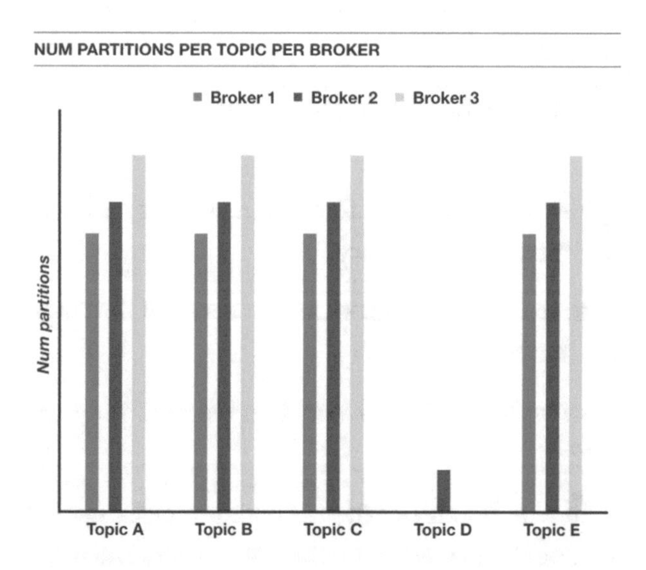


***Figure 5-14.*** *The partition that belonged to the topic with a single partition resided on the rogue broker*

To fix this production issue, we did two things. First, we added two more partitions to this topic. This helped balance the number of partitions across all brokers. Second, we got rid of some of the consumers for this topic. We found out that many of them didn't need to use that topic anymore. These two steps helped solve the issue.

This finding taught me a valuable lesson regarding partition skew—when checking for a partition skew among brokers, we need to check this skew not only in the scope of all the topics, but also in the scope of a single topic. As this production case shows, even a topic with a single partition that receives low traffic can cause a broker to become loaded.

This production issue can be summarized through its symptoms, potential causes, prevention measures, and monitoring options:

The symptom of this problem presents itself as a single broker with high CPU user time and high load average.

Potential causes for this issue might be that the broker was handling more incoming and/or outgoing traffic than other brokers. It could also be that this broker held a higher number of partitions (for all topics in the cluster) compared to other brokers. Additionally, this broker might host a topic that has a larger number of partitions on that broker, compared to the other brokers. Finally, it's possible that the broker hosted a topic with one or more of the following characteristics: many consumers, a high frequency of consumer requests, numerous producers, or a high frequency of producer requests.

To prevent such issues:

- Make sure to balance the number of partitions per topic across all brokers.
- Evaluate topics with a large number of consumers and consider whether all of them are necessary.
- For topics with many consumer requests, assess if the frequency of consumption can be reduced.
- For topics with many producers, consider whether they are all necessary.
- For topics with a high number of producer requests, see if their frequency can be decreased by increasing the batch size, linger time, or both.

Monitoring for this issue involves setting up alerts for the following scenarios: topics that their partitions are not balanced evenly among all brokers, topics with more than a certain number of consumers or consumer requests, and topics with more than a certain number of producers or producer requests.

# Summary

This chapter provided an in-depth exploration of CPU saturation in Kafka clusters, its causes, and potential solutions. It started by defining CPU saturation and a fully engaged CPU. Then it delved into the different types of CPU usage, including user time (`%us`), system time (`%sy`), I/O wait time (`%wa`), and software interrupts (`%si`). Each usage type was explained in the context of Kafka clusters and the potential reasons for their excessive utilization were discussed along with strategies for prevention and monitoring.

The chapter also emphasized the impact of log compaction on the CPU usage of the Kafka brokers and explained how this process, by retaining only the most recent value for each key, might require both RAM and CPU resources, potentially causing high resource utilization. A real production issue was presented where incorrect configurations for compacted topics resulted in system strain.

Lastly, the chapter discussed the impact of the number of consumers per topic on CPU usage, using a case study where a rogue broker caused lags and queuing.

Overall, this chapter serves as a comprehensive guide for understanding CPU utilization in Kafka clusters, helping you identify potential issues and implement effective strategies for prevention and monitoring.

The next chapter explores the important role of RAM in Kafka clusters. It explains why adding more RAM can be a critical improvement, sometimes even more so than adding CPU or disks, especially in reducing latency and boosting throughput. Cloud-based and on-premises solutions are compared, along with practical guides on how to monitor and manage RAM effectively. From understanding page cache to preventing system crashes, the next chapter gives you a comprehensive look at how RAM affects Kafka's performance, and how to optimize it.

## CHAPTER 6

# RAM Allocation in Kafka Clusters: Performance, Stability, and Optimization Strategies

This chapter explores the significance of RAM for the performance and stability of Kafka clusters, as well as its impact on consumers and producers. While a Kafka cluster relies on various hardware resources such as CPU, disks, and NIC, RAM holds a distinct position due to its influence on cluster stability and performance.

Many of the production issues that I've encountered were resolved and could have been prevented by simply adding more RAM to the cluster. This is because insufficient RAM can lead to disk saturation (resulting from high IOPS use) and increased CPU `sy%` and `wa%`. In both cases, consumers and producers experience delays.

The chapter begins by discussing the situations where adding RAM is more advantageous than focusing on CPU or disks. Subsequently, it elaborates on how Kafka interacts with the Linux page cache and why the page cache is so important to Kafka, followed by exploring the impact of inadequate RAM on disks. Additionally, we highlight the differences between adding RAM to cloud-based Kafka clusters versus on-premises solutions. Finally, the chapter provides a non-Kafka-related example that demonstrates the consequences of insufficient RAM in the page cache.




E. Eldor, *Kafka Troubleshooting in Production*, https://doi.org/10.1007/978-1-4842-9490-1_6

# Adding RAM to a Kafka Cluster

This section sheds light on why RAM isn't just another hardware component but a strategic resource in a Kafka cluster's operation. It explores how adding RAM can be a more effective solution than adding CPU cores or disks in some cases, and how this strategy varies between cloud and on-premises environments. With practical examples of how RAM affects performance, latency, throughput, and disk I/O, we'll unravel the complexities of making informed decisions about resource provisioning in Kafka clusters.

## The Strategic Role of RAM Over CPU and Disks

RAM distinguishes itself from other hardware resources in a Kafka cluster due to its unique effectiveness in terms of over-provisioning. To illustrate this point, consider the following example.

Imagine you have a Kafka cluster that currently possesses sufficient CPU capacity and disk storage. You plan on introducing additional producers and consumers to this cluster. While you know that the cluster doesn't require more CPU cores or disks to accommodate these new clients, the question arises: will more RAM benefit the consumers and producers?

This question is similar to asking whether the introduction of more consumers and producers would cause the brokers to read more data from the disks instead of utilizing the page cache. The true answer to this question can only be determined once these new consumers and producers are added to the cluster. At that point, you can evaluate whether additional RAM can prevent consumers from reading data directly from the disks. Monitoring disk reads/sec or employing tools like `cachestat`, which measure the miss and hit rates from the page cache, can assist in this analysis.

If you observe high disk utilization percentages due to increased read IOPS on the disks, the solution won't involve adding more disks but rather augmenting the RAM. This distinction is one of the reasons why RAM differs from CPU cores and disks. Insufficient RAM can lead you to misinterpret the climbing metrics and mistakenly conclude that you need to either add more CPU cores (due to elevated total CPU usage resulting from higher CPU system or wait time percentages) or add more disks (due to increased disk utilization caused by higher disk IOPS resulting from more reads from the disks).

A frequent dilemma faced by KafkaOps, especially when dealing with increased traffic or the need to accommodate more consumers and producers, is whether to scale the Kafka cluster up or out. While it may be tempting to say yes to one or both of these options, doing so can lead to over-provisioning of the Kafka cluster, unnecessarily driving up costs. In my experience, if the page cache is already highly utilized with the current traffic, scaling up the brokers by adding more RAM to each of them can often be more advantageous. Not only does this approach tend to address the immediate need, but it also typically offers a better return on investment compared to other scaling alternatives that might involve adding more disks or cores.

# Cloud vs. On-Prem RAM Expansion: Considerations and Constraints

There's a major difference between adding more RAM to machines in clusters that run on cloud versus on-prem. This section discusses the implications of adding RAM to the cloud versus to on-prem clusters.

## Adding RAM to the Cloud

In order to increase the amount of RAM for an instance in GCP or AWS, you need to switch to a new instance type that has more RAM and also different amounts of CPU cores, disk storage, and disk IOPS. If the chosen instance type has less CPU or disk space, you could get into CPU or disk space saturation, in which case you'll have to switch to an instance type with more RAM and also more CPU and/or disk space.

The same goes for switching to instance types with more RAM but also more CPU cores and disks—if you don't need them, you're just paying for resources that you don't need in order to get a resource that you do need (RAM).

It's important to choose the instance type that has the amount of RAM that you need while having the minimal amount of required CPU, disk space and IOPS. That's not always feasible in the cloud, which is why sometimes you can end up with a cluster that has the required amount of RAM but also more CPU and/or disks than your Kafka cluster really needs.

From a maintenance perspective, the process of switching instance types can also result in a temporary disruption of service as instances are swapped out.

## Adding RAM to On-Prem Kafka Clusters

Adding more RAM to an on-prem cluster is a different story than doing so on the cloud, since you have the freedom to add more RAM without adding other resources. However, there are limitations to the maximum amount of RAM that a machine can host, which depends on many factors (motherboard, processor, OS version, etc.).

In order to increase the amount of RAM in the broker machine, you need to consider not only the maximal amount of RAM it can host but also the number of memory slots in the machine, which determines the number of DIMMs the machine can contain. Each DIMM can be between 4-128GB RAM. So, for example, you could decide whether all the DIMMS will be in 16GB, 32GB, or more.

From a maintenance perspective, it's important to note that while on-premises systems offer greater control over hardware, they also require more manual management. You need to purchase the correct type of RAM, ensure compatibility, physically install the RAM, and potentially upgrade other components like the power supply.

Note that the maximum amount of RAM a machine can host is much higher in an on-prem machine than what's available in standard cloud instances. This could make on-premises systems more attractive options for certain high-memory workloads.

## Enhancing Kafka's Performance: The Benefits of Increasing Broker RAM

The allocation of additional RAM to a Kafka broker can yield several beneficial outcomes, enhancing the overall performance of your Kafka cluster. The following sections discuss those benefits.

### Performance Boost

When a Kafka broker is allocated with more RAM, it can make more effective use of the Linux page cache by keeping a greater portion of data in memory. This minimizes the requirement for frequent disk access, thereby accelerating both read and write operations, which in turn enhances the overall performance.

### Disk I/O Reduction

A larger amount of RAM allows Kafka to keep more messages in memory. Consequently, the frequency of disk I/O operations decreases, reducing disk latency. This can be particularly useful for alleviating potential bottlenecks in situations of high consumption or high throughput.

### Throughput Enhancement

By augmenting the available RAM, Kafka brokers can handle larger workloads from both consumers and producers. They can also process a higher number of messages concurrently. This leads to an improvement in throughput, enabling the broker to manage higher volumes of data more efficiently and support an increased number of consumers and producers.

### Latency Reduction

With increased RAM capacity, Kafka brokers can more frequently serve read requests directly from memory. This reduces the time it takes for consumers to access messages, leading to lower latency and quicker data retrieval. As a result, consumers experience a more responsive Kafka cluster.

## Understanding the Linux Page Cache

In Kafka's architecture, the Linux page cache plays a pivotal role in boosting both read and write operations. This section explores how Kafka utilizes this in-memory cache to enhance performance, while also considering the inherent tradeoffs between speed and fault tolerance. It examines the process of data writing in Kafka, how the page cache contributes to efficiency, and introduces tools like `cachestat` for monitoring cache utilization. Understanding these dynamics is crucial for optimizing Kafka's performance and reliability.

## Page Cache in Kafka: Accelerating Writes and Reads

This section delves into the importance of the Linux page cache to the performance of Kafka. One of Kafka's key performance optimizations lies in its efficient use of the Linux page cache, which serves as an in-memory cache provided by the operating system. This section explores how Kafka leverages this cache for improved performance and delves into the two-step process that Kafka employs when data is written into it. Furthermore, this section also discusses how this mechanism contributes to Kafka's speed and scalability, and covers the necessary precautions to ensure data integrity and fault-tolerance.

When data is written to Kafka, it goes through a two-step process. First, the data is written to the page cache, which acts as a cache layer in RAM. The page cache is a component of the operating system's memory management subsystem; it stores frequently accessed data from files. In this case, the data being written are the Kafka messages.

Once the data is in the page cache, it is marked as "dirty" because it has been modified and is yet to be persisted to the underlying storage (such as the disk). The page cache allows Kafka to achieve high write performance since it writes data to the cache in memory without the need for synchronous disk I/O operations. This asynchronous behavior provides a significant speed advantage. By leveraging the Linux page cache, Kafka achieves high write performance by temporarily storing data in memory before persisting it to disk.

Beyond write performance, Kafka's use of the page cache also improves read performance. When consumers read data from Kafka, the data is initially fetched from the disk and stored in the page cache, unless the data already existed in the page cache. Subsequent reads of the same data can be served directly from the page cache, significantly reducing disk I/O and improving read performance.

Figure 6-1 illustrates the flow of pages in the Kafka system, detailing how they are written by the producers and subsequently read by the consumers.

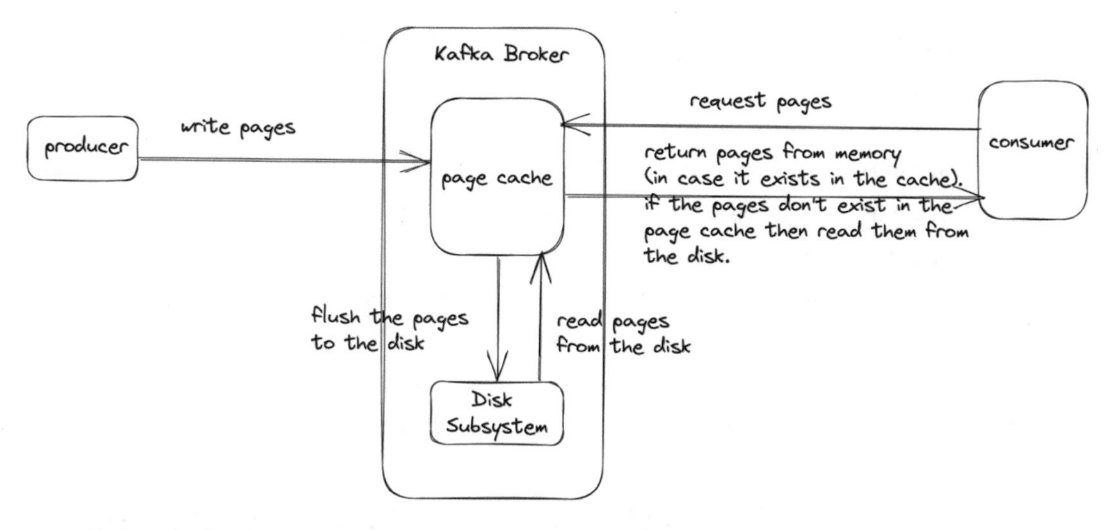


***Figure 6-1.*** *The flow of pages in the Kafka system*

As Figure 6-1 illustrates, when producers write messages to a Kafka broker, the pages containing these messages are placed into the page cache before being flushed to the disk. When consumers request specific messages from the broker, the system first searches for the corresponding pages in the page cache. If found, they are sent directly to the consumer; if not, the page cache retrieves the pages from the disk and forwards them to the consumer. These pages may remain in the page cache for some time, depending on the cache policy and access patterns. If other consumers subsequently request the same pages, they can be served directly from the page cache, improving efficiency by avoiding repeated disk reads.

However, Kafka's reliance on the page cache also comes with considerations. Over-reliance on the page cache without adequate memory can lead to excessive page swapping, causing a degradation in performance. Therefore, careful capacity planning and resource allocation are necessary to maintain optimal Kafka performance.

# Balancing Performance and Reliability: Kafka's Page Cache Utilization

It's important to note that the data residing in the page cache might not have been written to the underlying storage immediately. This introduces a tradeoff between performance and fault-tolerance. While Kafka benefits from faster writes due to the page cache, there is a slight risk of data loss if failures occur before the cache is flushed to disk.

To mitigate this risk, Kafka employs various strategies such as replication and acknowledgment mechanisms. Replication ensures that data is replicated across multiple Kafka brokers, providing fault-tolerance and data redundancy. The acknowledgment mechanisms, such as "acks" configuration, allow producers to receive confirmation of successful writes before considering them committed.

In the event of a failure where all Kafka replicas fail simultaneously, even with acks set to "all," there is still a chance of losing updates. This is because the page cache may not have sufficient time to persist the changes to the underlying storage before the failure occurs. Therefore, it's crucial to design Kafka deployments with fault-tolerance considerations and appropriate replication factors to minimize data loss risks.

## Monitoring Page Cache Usage Using the Cachestat Tool

As stated before, the Linux page cache plays a crucial role in Kafka's performance, as Kafka relies heavily on it to cache data and reduce disk I/O. Monitoring the page cache hit and miss ratio can provide valuable insight into how effectively your Kafka cluster is utilizing the page cache. A high hit ratio indicates that most read requests are being served from the cache (fast), while a high miss ratio indicates that many read requests have to go to the disk (slow).

`cachestat`, which is part of the Perf-tools suite developed by Brendan Gregg, is a powerful tool for monitoring cache usage on Linux. It provides real-time statistics on cache hits, misses, and hit ratio, and you can use it to monitor these performance metrics on the Kafka cluster.

The output columns of `cachestat` are as follows:

- `HITS`: The number of cache hits (read requests served from the page cache).
- `MISSES`: The number of cache misses (read requests that had to go to the disk).
- `DIRTIES`: The number of dirty pages (pages that have been modified and need to be written to disk).
- `READ_HIT%`: The cache hit ratio for read requests.
- `WRITE_HIT%`: The cache hit ratio for write requests.

If you notice one or more of the following—high cache miss ratio, lower hit ratio, or high number of dirty pages—it could indicate that your Kafka brokers are not effectively utilizing the page cache. This could lead to increased disk I/O and potential latency issues, as read and write requests have to go to disk instead of being served from the cache.

Possible causes for a high cache miss ratio include a lack of available RAM, a high rate of data eviction from the cache (possibly due to other memory-intensive processes running on the same machine), or consumers trying to read data that is not in the cache (possibly due to a high consumer lag).

A lower hit ratio in `cachestat` also indicates that a higher proportion of read requests are missing the cache and therefore need to be fetched from disk. This can lead to higher disk utilization because your storage subsystem has to handle more I/O operations.

You can monitor disk utilization using tools like `iostat` and `sar`. If you observe that a decrease in cache hit ratio correlates to an increase in disk utilization, it suggests that your Kafka brokers cannot effectively utilize the page cache, forcing them to rely more heavily on disk I/O. This can lead to increased latency and reduced performance, especially if your disks cannot handle the increased I/O load.

To conclude, monitoring your cache usage with `cachestat` can help you diagnose these issues, and adjusting your Kafka and system configurations based on these insights can help improve your Kafka cluster's performance.

# Lack of RAM and its Effect on Disks

When there is a lack of sufficient RAM for Kafka brokers, it will result in increased disk I/O, as Kafka cannot keep as much data in the page cache. This will make the performance of Kafka heavily reliant on the speed and configuration of your disks.

Figure 6-2 shows the output of the `top` command on one of the brokers of a Kafka cluster that suffers from lack of RAM in the machines.

```
%Cpu(s):  1.3 us,  0.3 sy,  0.0 ni, 78.0 id, 20.3 wa,  0.0 hi,  0.1 si,  0.0 st

MiB Mem :  8192.0 total,   121.2 free,  6930.5 used,  1140.3 buff/cache

PID USER     PR  NI   VIRT        RES     SHR  S  %CPU  %MEM    TIME+ COMMAND

43455 kafka  20   0   7863580     5.710g  8628 D  50.0   71.2    4:21.37 java
```

***Figure 6-2.** The output of the top command shows a broker that has a high I/O wait time because it reads data from the disks instead of from the RAM, probably due to lack of RAM*

In this example, the process of the Kafka broker uses about 5.7GB of memory, which is roughly 71.2 percent of the total available memory (8GB). The `wa%` value is 20.3 percent, which is quite high. This indicates that the CPU is spending a significant amount of time waiting for I/O operations to complete. This is likely because the Kafka broker has to read data from disk, due to insufficient RAM for the OS (in this case, ±2.3 GB of available RAM) to keep all the necessary data in the page cache.

The next section looks at how to optimize the disks in Kafka in order to allow the brokers to better handle a lack of RAM.

# Optimize Kafka Disks When the Cluster Lacks RAM

When operating a Kafka cluster with limited RAM, it's essential to implement strategic disk optimizations to maintain performance. By carefully choosing the right disk types, distributing logs, tuning the OS scheduling algorithm, and adjusting Kafka's specific policies, you can mitigate potential performance bottlenecks. The following sections explain some key recommendations to enhance Kafka's efficiency in a constrained RAM environment.

## Use SSDs Instead of HDDs

Solid State Drives (SSDs) are significantly faster than traditional Hard Disk Drives (HDDs) and can handle high rates of I/O requests, which is critical for Kafka's performance when RAM is insufficient.

## Distribute Logs Across Disks

Kafka enables the distribution of logs across multiple disks, enhancing I/O performance by spreading the load evenly. Through the use of the `log.dirs` configuration property, multiple directories can be specified, each residing on a separate disk. How Kafka brokers spread the log segments across these disks is influenced by whether the disks are configured using RAID or JBOD. A subsequent chapter delves into this configuration, detailing its effects on the distribution of files and on the overall disk performance.

## Tune OS Disk Scheduling Algorithm

On Linux, the I/O scheduler may be set to a default value that is not optimal for Kafka's access patterns. The `deadline` or `noop` schedulers are often better choices than the `cfq` scheduler for Kafka.

## Adjust Kafka's Disk Flush Policies

Kafka has several settings that control how often data is flushed to disk, including `log.flush.interval.messages` and `log.flush.interval.ms`. By adjusting these settings, you can reach a balance between durability and performance.

## Enable Log Compression

To reduce the amount of disk I/O, Kafka supports compressing message batches with different codecs (like Gzip, Snappy, LZ4, and Zstandard). This will reduce the amount of data written to and read from the disk.

## Enable OS Page Cache

Even if RAM is limited, ensure that you leave enough room for the OS page cache, as this can significantly reduce disk I/O. This can be done by verifying that the only process that runs on the broker that isn't an OS process is the Kafka process itself.

## Monitor Disk Usage and I/O

Regularly monitor the disk utilization and I/O operations to detect and troubleshoot performance issues. Tools like `iostat`, `vmstat`, and `dstat` can be helpful.

# A Lack of RAM Can Cause Disks to Reach IOPS Saturation

In order to show the effect of a lack of RAM on the disks of the Kafka brokers, this section looks at a Kafka cluster in which some of its consumers were lagging. The root cause for that lag turned out to be a lack of RAM in the Kafka clusters.

Figure 6-3 shows the consumer lag; all the consumers are lagging behind in all partitions.

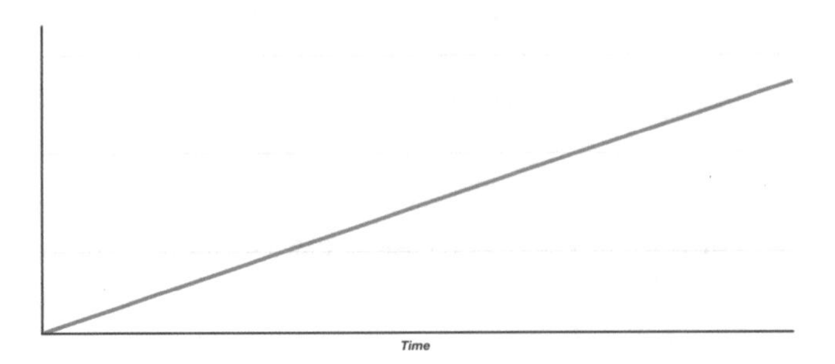


***Figure 6-3.*** *The sum of lag (in millions) in all the consumers over all the partitions over time. As time goes by, the lag continues to increase*

When the consumer lag started to increase, you could see the following behaviors.

Figure 6-4 shows that the throughput of the reads from the Kafka disks increased until they reached the disks' maximal throughput.

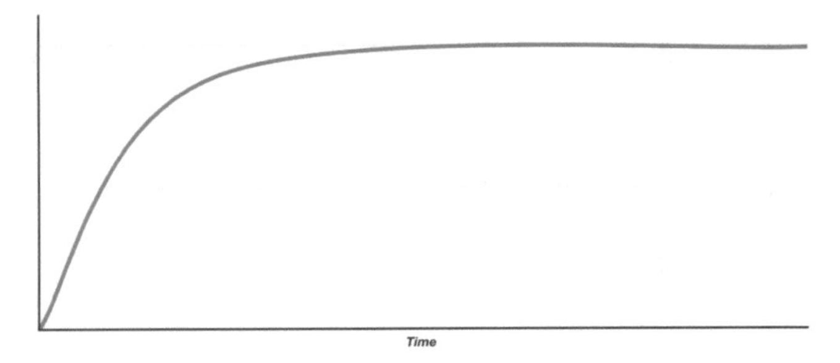


***Figure 6-4.*** *The throughput of reads from disks in MB (rMB/s)*

Figure 6-5 shows that the IOPS of reads from the disks also increased until they reached the maximal IOPS that the disks provide.

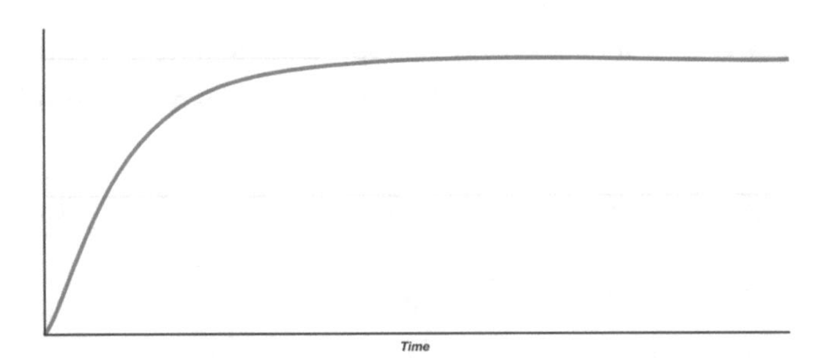


***Figure 6-5.*** *The read IOPS from the disks (r/s)*

Figure 6-6 shows that the CPU I/O wait time increased until it stabilized at 10 percent, which is pretty high for I/O wait.

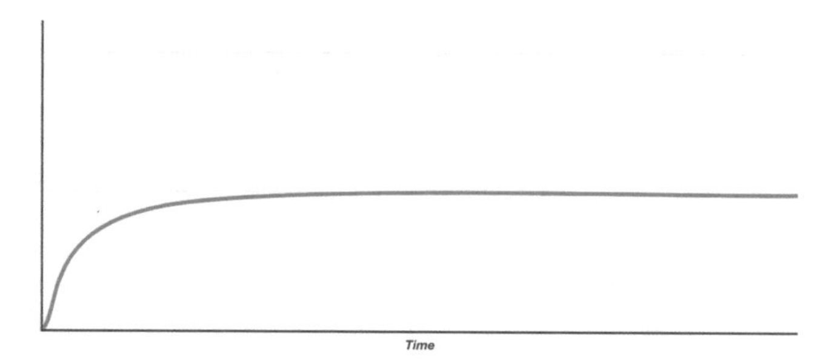


***Figure 6-6.** The wait time increased due to latency in the disks*

When you look at the disk utilization and compare it to the page cache hit ratio, you'll see a negative correlation—the lower the page cache hit ratio, the higher the disk utilization. See Figure 6-7.

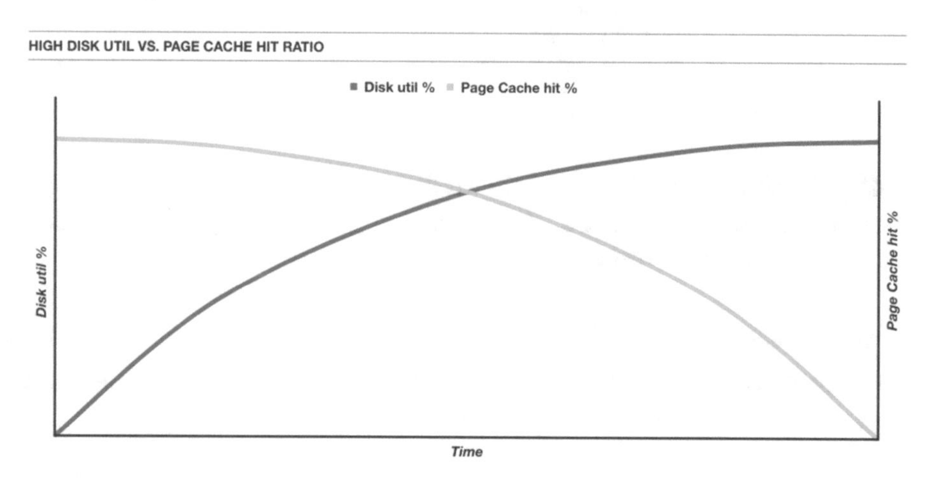


***Figure 6-7.** A negative correlation between hit ratio from the page cache and the disk utilization—as the hit ratio goes down, the disk utilization increases until it reaches IOPS saturation*

To conclude, this example shows how a page cache hit ratio has an immediate effect on disk utilization. The higher the hit ratio, the lower the disk utilization and vice versa—the higher the miss ratio, the higher the disk utilization and consequently the higher the CPU wait time.

Here's another example of a Kafka cluster (this time it is on-prem) where its consumers suffered from recurring lags. We decided to triple the amount of RAM in that cluster and measured the disk utilization before, during, and after the addition of RAM.

Figure 6-8 shows the effect of adding RAM on the disk utilization of the disks in the cluster.

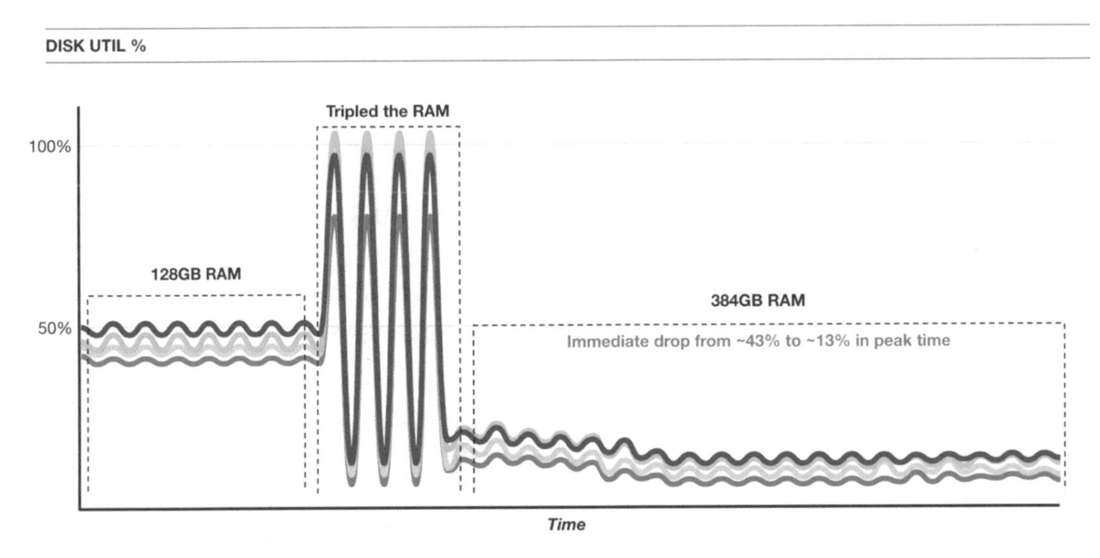


***Figure 6-8.*** *The disk utilization in a Kafka cluster that suffers from lack of RAM. There's a correlation between the amount of RAM in the brokers and a reduction in the disk utilization*

Note that the disk utilization dropped linearly based on the amount of RAM added—from 43 to 13 percent. This is another indication that the cluster lacked RAM, and that lack of RAM and high disk utilization go hand in hand.

## Optimize Kafka in Terms of RAM Allocation

Optimizing Kafka's performance in terms of RAM allocation requires careful adjustments to the operating system, the JVM settings, and the Kafka configuration itself. The following sections discuss some recommendations.

## Set vm.swappiness to the Minimum Possible Value

The `vm.swappiness` parameter determines how aggressively the kernel will swap memory pages versus dropping pages from the page cache. A higher value increases swap aggressiveness, while a lower value tells the kernel to swap as little as possible.

For Kafka, where high I/O performance is critical, it is typically recommended to minimize swapping. Swapping can cause Kafka to pause, leading to high latency and potential timeouts. For example, if a Kafka broker is experiencing high latency and frequent pauses, and upon investigation, `vm.swappiness` is found to be set to a high value, lowering the value will reduce these issues.

Similarly, minimizing or disabling swappiness on the ZooKeeper servers is crucial. Swapping on ZooKeeper machines can create latency, leading to delays in synchronization and potential inconsistencies in the distributed operations. This could affect the Kafka cluster's responsiveness, integrity, and reliability, emphasizing the need to control swappiness on both Kafka brokers and ZooKeeper servers.

## Increase the File Descriptor Limits

Kafka brokers maintain a file for each partition, and they have many connections to other brokers, producers, and consumers. Each open file or socket consumes a file descriptor, and there's a finite limit to how many a Kafka broker can open. If the file descriptor limit is too low, Kafka may experience errors leading to potential broker failure. For instance, the broker might start rejecting new client connections or fail to open new log segments if it reaches the file descriptor limit. This limit can be modified using the `ulimit` command.

## Increase the Limit of Memory-Mapped Files

In addition to the file descriptor limit, Kafka brokers also deal with a limitation on the number of memory-mapped files, controlled by the kernel parameter `vm.max_map_count`. The broker maps files into memory, and reaching this maximum count can lead to similar errors and broker failures as hitting the file descriptor limit.

## GIVE at Least 32GB RAM to Your Kafka Brokers

Kafka uses the page cache to buffer reads and writes to disk, and the more memory you can give to the page cache, the better your I/O performance will be. If your brokers have insufficient memory, they may struggle to keep up with write and read requests, leading to increased latency and lower throughput. For example, if a Kafka cluster is experiencing slow message consumption rates, increasing the amount of RAM allocated to each broker might help improve the consumption rates.

## Monitor Garbage Collection Times Closely

Kafka runs on the JVM, which uses garbage collection (GC) to free up memory. However, GC pauses can impact Kafka's performance. If GC is taking a significant amount of time, it may be necessary to adjust the JVM settings or upgrade to a newer JVM version with a more efficient garbage collector. If a Kafka broker is experiencing frequent full GC activity, it could lead to noticeable pauses in processing, resulting in increased message latency.

## Tuning JVM Options

The JVM settings can have a significant impact on Kafka's performance. For example, you might want to adjust the heap size settings (`-Xmx` and `-Xms`) to ensure Kafka has enough heap space, but not so much that it causes excessive GC pauses.

## Using Appropriate Instance Types When Deploying on a Cloud Platform

If you're deploying Kafka on a platform like AWS, choosing an instance type with a high memory-to-vCPU ratio can help ensure you're getting the most out of your RAM allocation.

## Balancing Topics and Partitions Across Brokers

To prevent any one broker from becoming a bottleneck, it's important to distribute topics and partitions evenly across the available brokers. This can help ensure that each broker's RAM is used efficiently.

## Dealing with Garbage Collection (GC) and Out-Of-Memory (OOM)

Like any system based on the Java platform, Kafka brokers are susceptible to the potential pitfalls of Java's garbage collection (GC) processes. Mismanaged or poorly tuned garbage collection can significantly impact Kafka's performance and reliability, undermining its ability to deliver messages promptly and reliably. As we delve deeper into this topic, we'll explore the intricate workings of garbage collection within Kafka brokers, its implications for Kafka's performance, and the potential consequences of high GC frequency or long GC pauses.

Kafka's performance hinges on its ability to manage memory effectively. The Kafka brokers utilize two key types of memory: the Java Virtual Machine (JVM) heap and the page cache. The JVM heap is used for transient data structures and for operations such as log compaction and replication of log segments in partitions. The page cache, on the other hand, is managed by the operating system and is used to cache Kafka log files for faster access.

However, when the JVM heap becomes too populated, the garbage collector steps in to clear unused objects, a process that can be quite resource-intensive. High GC frequency or long GC pauses can significantly impact Kafka in several ways, discussed next.

## Latency Spikes

When the garbage collector kicks in, it can cause temporary pauses in Kafka's processing. This GC pause can result in spikes in latency, negatively impacting the performance of the cluster and disrupting the smooth flow of message processing for both producers and consumers.

## Resource Utilization

The garbage collection process is CPU-intensive. High GC activity can consume significant CPU resources, limiting what's available for other processes. This, in turn, can degrade overall cluster performance and affect other services running on the same nodes.

## System Stability

GC activity is directly tied to system stability. If the garbage collector struggles to reclaim memory in a timely manner, it can lead to JVM out-of-memory errors. Such situations can cause Kafka brokers to crash, leading to potential service disruptions, data loss, and a cascading effect of rebalancing across the remaining brokers.

## Impact on ZooKeeper Heartbeat

Long GC pauses can cause Kafka brokers to miss their ZooKeeper heartbeat, resulting in the broker being marked as dead. This can trigger unnecessary partition rebalancing, causing additional load and reducing the overall throughput of the system.

Therefore, it's crucial to monitor GC performance and tune the Kafka brokers to minimize the impact of GC on the system's performance. Effective GC tuning and proper memory management can help maintain Kafka's high performance and its promise of real-time data delivery.

# Measuring Kafka Memory Usage

This section discusses the Linux commands that measure the memory usage of the Kafka machine and Kafka process. Utilizing the appropriate Linux commands can provide a comprehensive view of how memory is being utilized, both at the machine level and specifically by the Kafka process itself. This understanding is crucial in diagnosing issues, planning capacity, and tuning the operating system and Kafka configurations for optimal performance.

The following list explains some valuable commands and methodologies that you can employ to analyze and interpret the memory consumption of your Kafka machine and Kafka process.

- `top`: Run `top` and look for the Kafka process. Check the RES (resident memory size) column to get an idea of how much memory the Kafka process is using. If you know the exact name of the Kafka process, you can use `top -p $(pgrep -d',' -f kafka)` to filter the `top` output specifically for Kafka.

- `vmstat`: Run `vmstat -s` to get a snapshot of various aspects of your system's memory usage. Look for "used memory," "free memory," "buffer memory," and "swap memory" to understand how memory is being used on your system.
- `free`: Run `free -h` to see a summary of your system's memory usage in an easy-to-read format. This includes total, used, free, shared, buff/cache, and available memory.
- `pidstat`: Since with `pgrep` you get other PIDs that matched the Kafka world, first find the PID (process ID) of your Kafka process using `pgrep -f kafka`. Then run `pidstat -r -p [PID] 1` to monitor the Kafka process's memory usage at one-second intervals.
- `sar`: `sar -r 1` will provide system memory usage stats (including RAM and swap) every second. If you have historical data enabled for `sar`, you can use `sar -r -f /var/log/sa/sa[day]` to view memory usage stats for a specific day.
- `dstat`: Run `dstat -g -m 1` to see stats about memory usage and garbage collection at one-second intervals.
- `jstat`: If you're specifically interested in JVM memory usage for Kafka you can use `jstat`. First, get the PID of the Kafka process using `pgrep -f kafka`, then run `jstat -gc [PID] 1s` to get garbage-collected heap memory details at one-second intervals.

---

**Note** Remember to replace `[PID]` with the actual process ID of your Kafka process.

---

# The Crucial Role of RAM: Lessons from a Non-Kafka Cluster

The importance of RAM cannot be overstated. This was vividly illustrated when a cluster running an application written in Go language encountered a crippling issue.

The application would gradually consume all available RAM on a machine in the cluster. Interestingly, each occurrence affected a different machine, leading to a game of "musical chairs" that ultimately caused the cluster to cease functioning. It was not because of high CPU usage or disk utilization; both metrics were surprisingly low. Instead, a high percentage of system CPU time (CPU `sy%`) became the telltale sign of trouble.

We finally concluded that each machine's "death" (instances where AWS killed the machine) was directly linked to an available RAM count of zero. A single process was monopolizing most of the CPU time, but it was primarily wait time (`wa%`) rather than user time (`us%`).

Contrary to initial suspicions, this was not a garbage collection (GC) issue, as GC typically consumes user CPU time, not wait time. It's crucial to note that this process had nearly all the machine's RAM allocated to it, which could potentially lead to misses in the page cache.

CPU wait time (`wa%`) signifies the time the CPU is waiting for device I/O, which could involve block devices (disk) or network devices. In this instance, the disk seemed to be the bottleneck.

The machine was attempting to process an enormous volume of disk reads, reaching 200MB/sec. This high level of disk activity might have been due to the lack of available RAM for the page cache. As the process was relentlessly reading from the disks, the disks reached their maximum capacity (100% utilization), and the I/O wait time soared to 50%. This was the root cause of this latency issue.

We contemplated several potential solutions:

- Allocate less RAM to the Go process (less than 14GB). Then, assess if the disk reads drop below 200MB/sec, and if this reduces the disk utilization and I/O wait time.
- Use a machine with more RAM, but keep the Go process's heap size consistent with the current configuration. This would ensure more available RAM than the current situation, which is zero. Then, verify if the disk utilization falls below 100% and if the I/O wait time is significantly reduced.
- Use a machine with a disk that offers more I/O operations per second (IOPS) and the same RAM.

Since we were operating in an AWS environment, a machine with more RAM would likely also provide more disk IOPS. Therefore, we decided to scale up the machine but made sure that the Go process didn't consume more than half of the machine's RAM. This experience underlines the critical role of RAM in maintaining system performance and stability.

# Summary

This chapter delved into the significant effects of RAM on Kafka clusters. It elaborated on why adding RAM can sometimes be more crucial than adding CPU or disks, as it reduces latency and improves throughput. The chapter further differentiated the impacts of adding RAM in cloud-based machines versus on-premises ones, emphasizing the scalability and cost-effectiveness of cloud solutions.

We then outlined the positive outcomes of adding more RAM to Kafka brokers, predominantly enhancing the brokers' performance and increasing the page cache. We shed light on how the page cache significantly impacts Kafka's read and write operations, with a larger cache leading to faster data retrieval and insertion. However, we also emphasized that this performance boost needs to be balanced with fault tolerance considerations.

We provided a practical guide to monitoring the page cache usage with the `cachestat` tool, enabling effective control over Kafka's performance. We also emphasized the detrimental impact of a RAM shortage on disks, potentially causing the disks to reach IOPS saturation. In such a situation, I recommend optimizing Kafka disks to counteract the lack of RAM.

We then learned how to optimize Kafka in terms of RAM allocation, ensuring that the Kafka cluster runs smoothly and efficiently. We further explored the relationship between garbage collection (GC), out-of-memory (OOM) errors, and RAM, highlighting the importance of proper memory management to prevent system crashes.

We provided a guide to Linux commands that measure the memory usage of the Kafka machine and the Kafka process, enabling KafkaOps to monitor and manage RAM usage effectively. Lastly, we took a detour to read about the role of RAM in non-Kafka clusters, underlining the universal importance of RAM in all types of clusters.

In summary, this chapter serves as a thorough guide to understanding the impact of RAM on Kafka clusters, offering practical solutions to common problems related to RAM allocation, management, and optimization.

The next chapter takes a closer look at disk I/O overload in Kafka clusters, focusing on how it affects consumers and producers. Starting with an introduction to key disk performance metrics, we'll dive into practical applications, such as diagnosing latency issues and detecting faulty brokers. Real-life examples highlight the significance of careful adjustment and monitoring of disk.io threads to ensure optimal performance. Whether you're interested in understanding how Kafka reads and writes to disks or seeking to improve efficiency and reliability, the next chapter offers valuable insights into the intricate world of disk utilization in Kafka.

CHAPTER 7

# Disk I/O Overload in Kafka: Diagnosing and Overcoming Challenges

Disk I/O overload can create significant challenges in Kafka clusters, affecting both consumers and producers. Consumers may accumulate a backlog leading to data loss, while producers might buffer data until the buffer overflows, also resulting in loss of data. Recognizing whether these problems stem from disk I/O activity is vital to prevent data loss.

This chapter explores various scenarios that can cause the disks in a Kafka cluster to halt or delay the functioning of consumers and producers. It starts by outlining the relevant disk performance metrics for evaluating disk usage and then discusses how to determine if disk latency is affecting Kafka brokers, consumers, or producers.

Following that, the chapter delves into real-life production problems that were diagnosed using these metrics. This includes the detection of a faulty Kafka broker through these metrics, the consequences of having too many disk.io threads, the importance of checking disk performance during peak usage times, and the effect of disk.io threads on the performance of brokers, producers, and consumers.

Through these insights, the chapter aims to equip you with the understanding needed to identify and address disk-related issues in Kafka clusters, enhancing the efficiency and reliability of your systems.




E. Eldor, *Kafka Troubleshooting in Production*, https://doi.org/10.1007/978-1-4842-9490-1_7

## Disk Performance Metrics

In assessing potential disk I/O issues in your Kafka cluster, various disk performance metrics are indispensable. Most of these, except the last one, can be observed through the Linux `iostat -x` command.

- Among these metrics, *IOPS* (Input/Output Operations Per Second) provides insight into the number of read and write operations performed by a disk per second, serving as a gauge of the disk's input/output efficiency. This encompasses the number of read requests (r/s) and write requests (w/s) issued to the device per second.
- *Throughput* refers to the volume of data that can be read or written to a disk per second, reflecting the disk's data transfer rate. This can be broken into the number of megabytes read (rMB/s) and written (wMB/s) to the device each second.
- *IOPS utilization* signifies the proportion of time the disk is engaged in read and write operations, representing a measure of the disk's workload. This can be expressed as the percentage of CPU time (CPU `util%`) during which I/O requests are directed at the device.
- The *service time* of a disk quantifies the time required for the disk to complete a read or write operation, and it's indicative of the disk's latency or response time. For instance, if the disk utilization is 80 percent and IOPS is 400, the service time is calculated as 800 ms/400, or 2 ms.
- Another noteworthy metric is `%iowait`, though more related to applications. In Kafka, this value is typically attributed to disk latency rather than network devices. `%iowait` signifies the percentage of time the CPUs remained idle while awaiting a pending disk I/O request.

While these metrics are vital and effective for detecting latency induced by disks, they only represent a subset of available disk performance indicators. Other potentially helpful metrics, such as *disk queue length*, indicating the number of pending read and write requests, and *disk average response time*, denoting the average response duration of the disks, are also valuable but not detailed in this chapter. These chosen metrics, however, provide a robust foundation for identifying and diagnosing disk-related latency issues in Kafka clusters.

# Detecting Whether Disks Cause Latency in Kafka Brokers, Consumers, or Producers

When Kafka brokers' consumers or producers experience latency, investigating the disks can often help us to understand the root cause of that delay. Since disks frequently play a role in many latency issues, and sometimes even cause them, this section begins by explaining how Kafka utilizes its disks. Following that, we explore various symptoms that may lead to latency for consumers and/or producers, and learn how the previously described disk performance metrics can be leveraged to pinpoint the underlying causes.

# How Kafka Reads and Writes to the Disks

To determine if disk I/O performance is causing latency in Kafka brokers, consumers, or producers, we must first understand how Kafka interacts with its disks. The method Kafka uses to write to and read from the disks influences how disk utilization should be assessed. It's essential to distinguish between the write and read processes since they represent two different flows, and the operating system responds differently to each case.

## Writes

The writes to the filesystem are performed by producers, which send messages to the brokers, and by the brokers themselves. The brokers write messages that they have fetched from other brokers, and then receive from these same brokers as part of the message replication process.

The flow of the writes to the disks of a broker is shown in Figure 7-1.

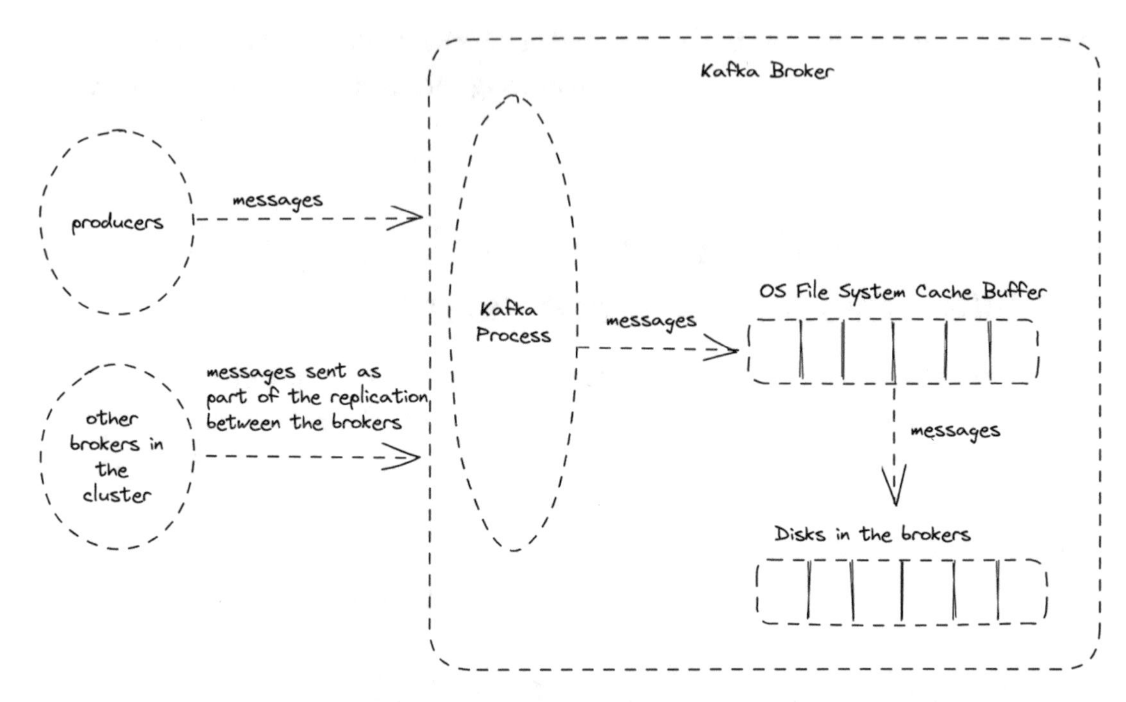


***Figure 7-1.*** *How messages are sent to the broker by producers or by other brokers within the cluster*

As shown in Figure 7-1, the Kafka threads receive these messages and relay them to the OS filesystem cache for buffering (it's important to note that Kafka itself doesn't buffer the messages). Then, the OS threads intermittently flush these messages to the disks in bursts, resulting in periods of no writes at all, followed by instances where disk utilization spikes dramatically.

The phenomenon of these bursts is crucial to understand, as it's a common misconception to equate high disk utilization in Kafka brokers with saturation. This view is flawed because the way that the OS flushes data to the disks creates periods where the disks in Kafka either rest (with low utilization) or work intensely (with high utilization) due to a burst of messages being flushed. Therefore, high utilization doesn't necessarily mean that the disks are saturated.

Disk utilization can be influenced by various factors related to disk writes, including high throughput from the producers, replication factor, number of segments per partition, size of each segment within a partition, and the values of `batch.size` and `linger.ms`.

## Reads

The reads from the filesystem are performed by the following:

- Consumers that fetch messages from the brokers
- Brokers that fetch messages from other brokers as part of the message replication

Disk reads are performed whenever data doesn't exist in the page cache, and since the reads aren't buffered by the OS, reads aren't performed in bursts.

The disk utilization is affected by several factors, which are related to disk reads:

- The number of consumers
- Any consumer lag: Consider the following case—a consumer reads 10K per batch, and the `max.poll.records` parameter is configured to 10K. The consumer stops for some reason, and after an hour, it's restarted. On that hour, there were 3.6M messages that weren't consumed, so now that the consumer starts, it will read 10K messages per second until the lag is over, which will take 3.6M/10K = 360 seconds just to consuming that lag. In these six minutes, the consumption lag will be the same as usual, but messages containing this data might be read from the disks because they don't exist in the page cache anymore.
- The amount of available RAM in the broker machine

# Disk Performance Detection

When encountering one or more of the following symptoms, one of the suspects can be disk I/O. The following sections list several symptoms, and for each symptom, we'll go over the disk I/O performance metrics that can be used to detect the cause.

## Data Skew in the Scope of a Single Broker

A disk with higher IOPS or throughput can mean that data isn't distributed well among all the disks in the broker.

Figure 7-2 shows a skew in the IOPS per disk in a specific broker.

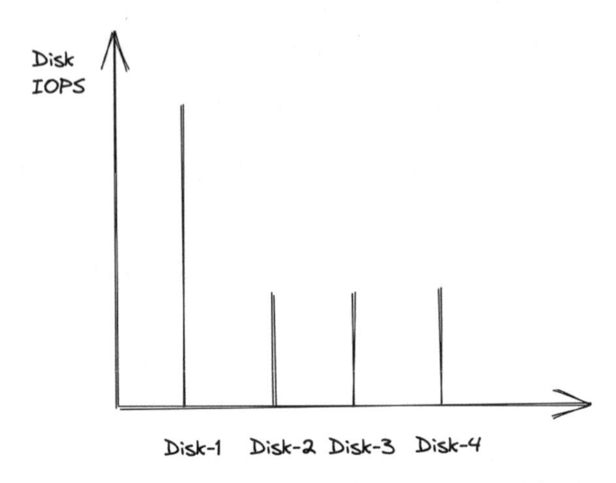


***Figure 7-2.*** *These are the IOPS (read and write operations/sec) per disk in all the disks of a specific broker. The broker has four disks and Disk-1 has higher IOPS compared to the others*

There can also be a skew in the throughput per disk in a specific broker, as shown in Figure 7-3.

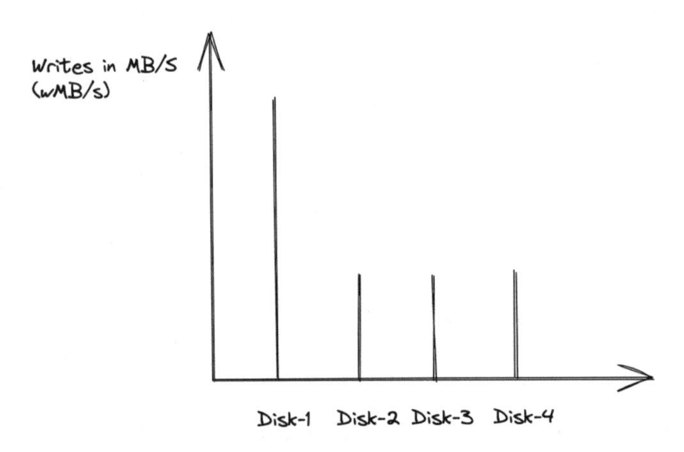


***Figure 7-3.*** *These are the write throughputs (wMB/s, or writes MB/sec) per disk of all the disks in a specific broker. The broker has four disks and Disk-1 has higher wMB/s compared to the others*

In such a situation, if the disk utilization percentage is significantly higher than the rest, it can lead to latency issues that may affect the replication of partitions residing on that disk, consumers consuming from those partitions, and producers writing to the partitions located on that disk.

## Data Skew in the Scope of a Kafka Cluster

If the average write throughput and IOPS utilization are higher in all the disks of one broker (compared to these metrics in the other brokers in the cluster), as shown in Figures 7-4 and 7-5, then this broker might receive more data than the other brokers.

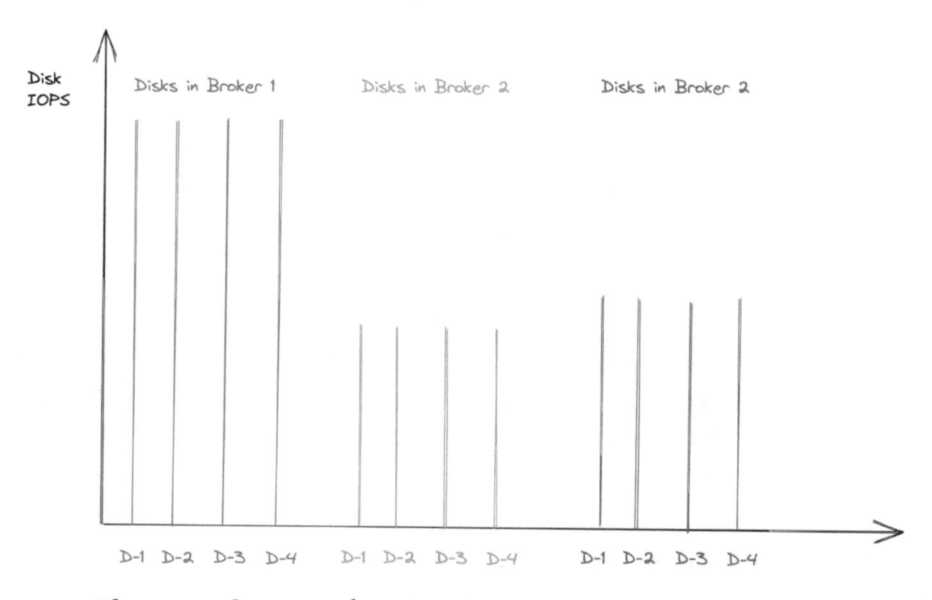


***Figure 7-4.*** *These are the IOPS (read and write operations/sec) per disk in all the disks of all the brokers in the cluster. There are three brokers in the cluster and each broker has four disks. The IOPS is twice as high in the disks of Broker-1 compared to the other brokers*

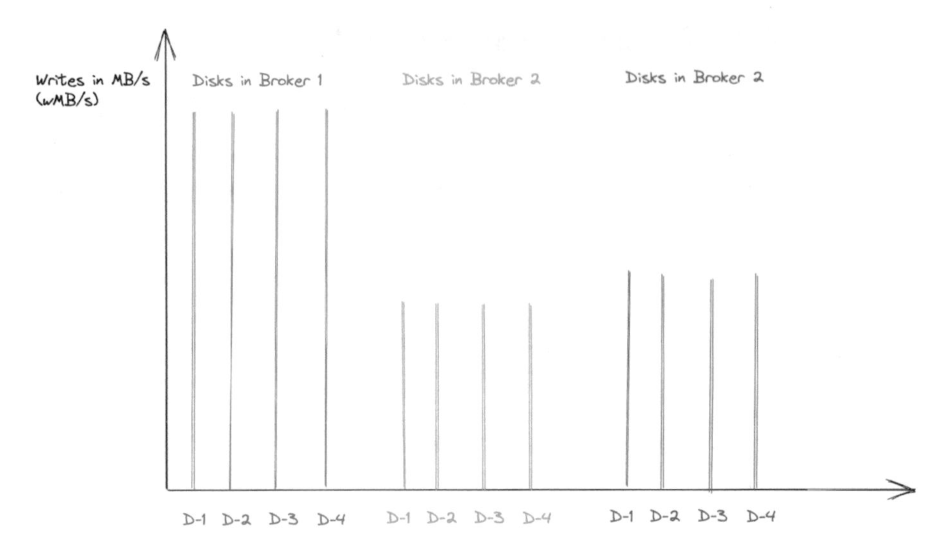


***Figure 7-5.*** *These are the write throughput (wMB/s, or writes MB/sec) per disk in all the disks of all the brokers in the cluster. There are three brokers in the cluster and each broker has four disks. The wMB/sec is twice as high in the disks of Broker-1 compared to the other brokers*

## Consumer Lag from a Specific Broker

If IOPS utilization is higher on the disks of one broker, this can be due to a consumer lag on partitions that reside on that broker. This lag can cause a higher IOPS count when some of these messages no longer exist in the page cache, which causes the OS kernel to fetch the data from the disks.

## Slow (Faulty) Disk

In the scenario where all disks have similar IOPS (input/output operations per second) and throughput, but one disk has a higher average utilization percentage than the others, this could signal a faulty disk, and it might be wise to consider replacing it. When dealing with a Kafka cluster deployed on the cloud, another possible explanation for a slow disk can be what's known as a noisy neighbor.

In the context of a Kafka broker's disk I/O, a *noisy neighbor* refers to another tenant or instance utilizing the same disk in a manner that hampers the broker's performance. Since they share the I/O subsystem and access the same disk simultaneously, noisy neighbors can create contention for available I/O resources, leading to a cascade of problems, including increased latency, reduced throughput, and a general decline in overall disk performance.

## Real Production Issue: Detecting a Faulty Broker Using Disk Performance Metrics

I encountered a production issue in which a broker with faulty disks caused an entire cluster to stop functioning—consumers started to lag and the buffers on the producers' side started to increase. During that time, the traffic didn't increase compared to every other day.

Then I saw the following disk I/O-related symptoms in the faulty broker (compared to the other brokers in the cluster):

The average read service time from a specific disk was higher, as shown in Figure 7-6.

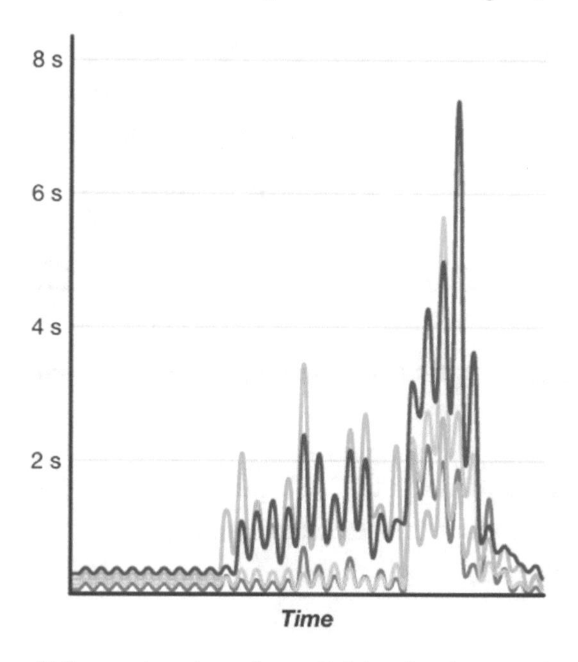


***Figure 7-6.*** *Average disk service time (per disk) is higher in the red disk compared to the other disks*

The average number of read bytes from that specific disk was also higher, as shown in Figure 7-7.

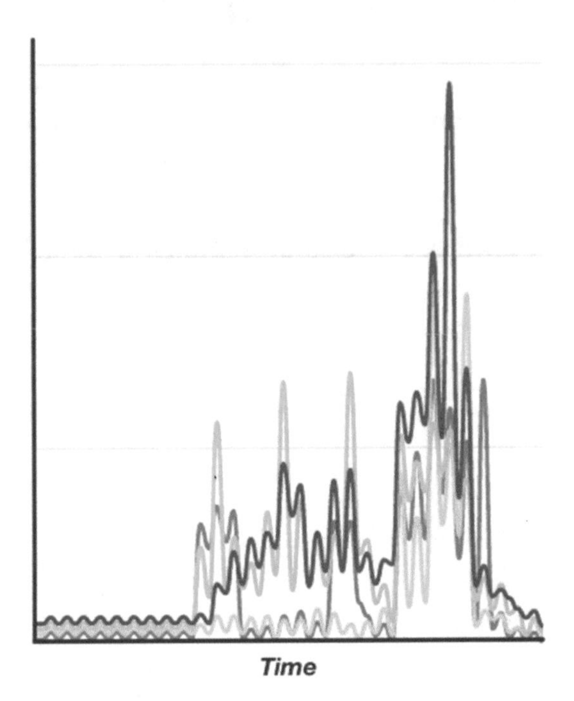


***Figure 7-7.*** *Disk I/O read bytes (per disk) is higher in the red disk compared to the other disks*

The size of the request queue on the broker that hosted that disk was higher (the request queue contains produce and fetch requests that are sent by clients to the brokers), as shown in Figure 7-8.

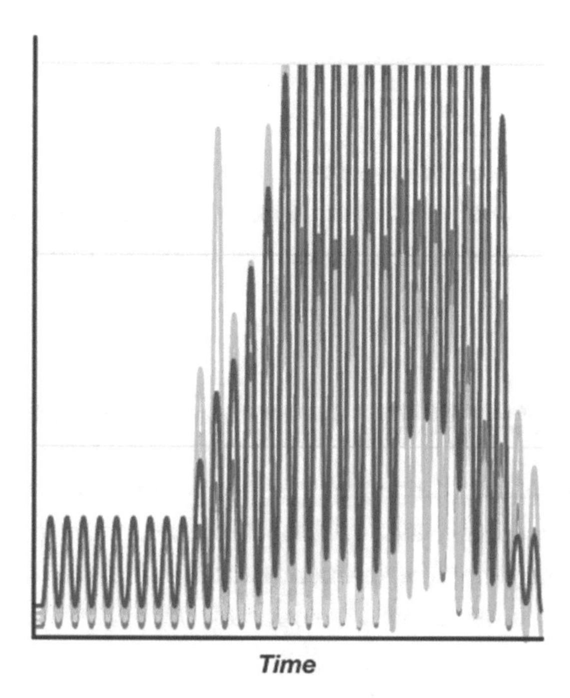


***Figure 7-8.** Request queue size (per broker) is higher in the broker that hosts the disk (the one with the high I/O read time and I/O read bytes) compared to the other disks*

The produce latency was also higher, as shown in Figure 7-9.

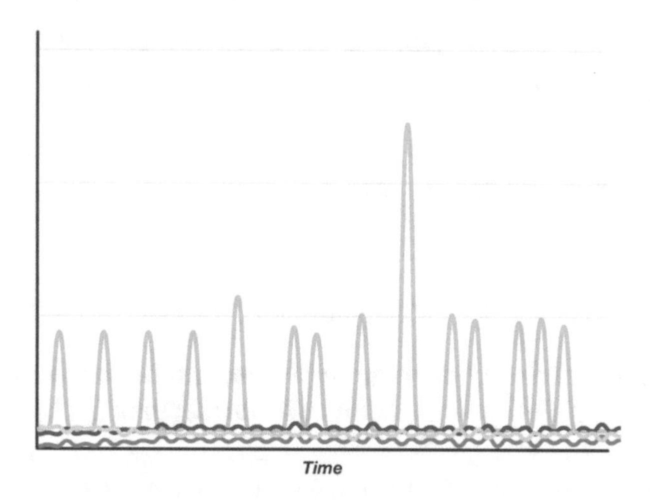


***Figure 7-9.** Produce latency 99th percentile (per broker) is higher in the broker that hosts the disk (the one with the high I/O read time and I/O read bytes) compared to the other disks*

## Discussion

This combination of disk and Kafka metrics helped us detect a faulty broker.

That broker had more reads from a single disk, which caused it to serve and consume requests slower compared to the other brokers. The size of the request queue that contains consume and produce requests was higher on that broker compared to the other brokers.

# The Effect of Too Many disk.io Threads

I/O threads take requests from the request queue and process them. The number of I/O threads is determined by the `num.io.threads` configuration parameter. In order to improve throughput, the general recommendation is to increase their number up to the number of disks. From my experience, the optimal `num.io.threads` is twice or three times the number of disks.

Figure 7-10 shows a cluster in which the normalized load average in all brokers started to rise during peak times and reached above 1.0.

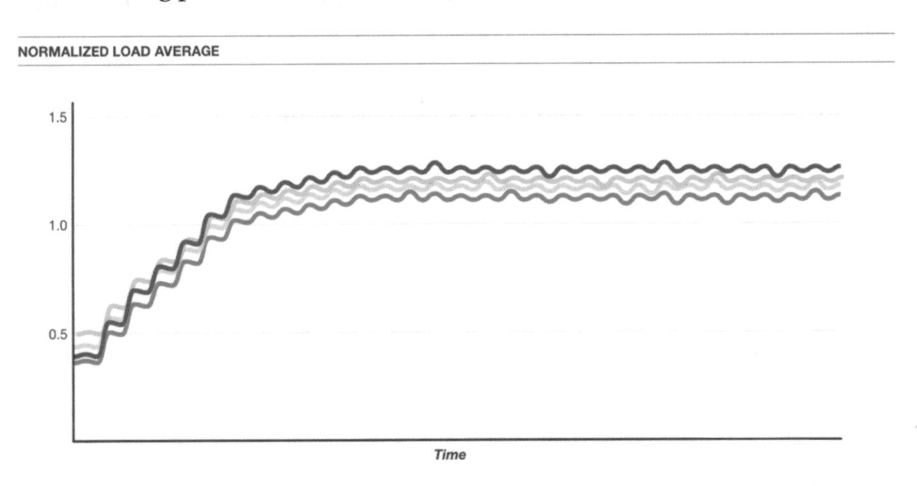


***Figure 7-10.*** *Normalized load average in all brokers started to climb and reached ±1.3, which indicates there's a saturation in some resource in the cluster*

The problem was that we didn't know which change caused the load average to grow, so we added new brokers to check whether they suffer from the same symptom, and spread the partitions evenly across the old and new brokers.

Surprisingly, the load average of the new brokers was lower than the load average of the old brokers, as shown in Figure 7-11.

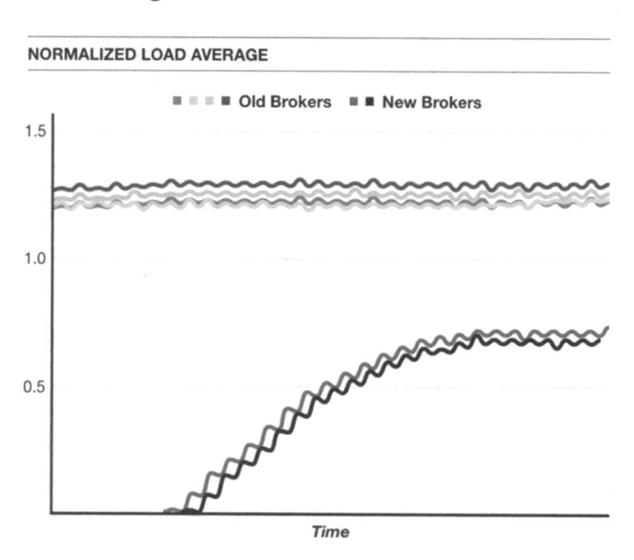


***Figure 7-11.** Normalized load average in the old brokers was ±1.3, while in the new brokers it reached ±0.7. So while the old brokers remained saturated, the new brokers didn't*

When looking for a reason that the load average on the old brokers was high, we noticed that the server interruptions rate (which is the CPU `%si` shown in the `top` command) was almost twice that in the old brokers, as shown in Figure 7-12. This correlates with the normalized load average of ±1.3 in the old brokers compared to 0.7 in the new brokers.

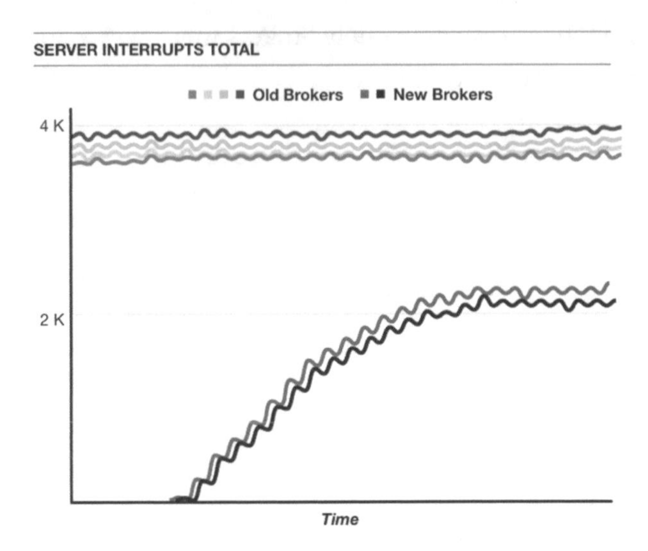


***Figure 7-12.** Server interruptions rate (which is shown as %si in the top command) reached ±4K in the old brokers, while in the new brokers it was 2K*

Since the server interruptions rate is related to the number of threads, we started looking for potential suspects and the number of I/O threads was one of these suspects.

Once we reduced the number of I/O threads from 6 to 2, the load average returned to normal in the old brokers, as shown in Figure 7-13.

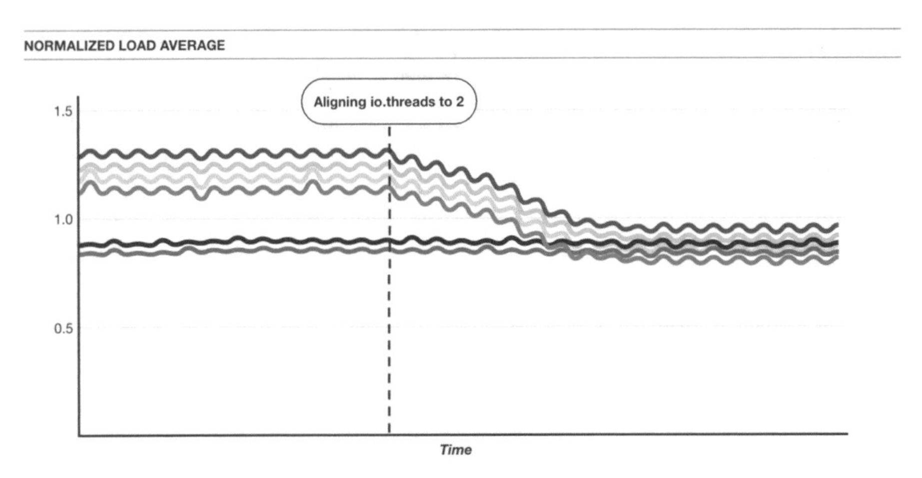


***Figure 7-13.** Once the number of I/O threads was reduced from 6 to 2 in the old brokers, the normalized load average in the old brokers was reduced and became almost the same as the normalized load average in the new brokers*

After that we removed the new brokers from the cluster, because the only reason we added them was to compare the configuration between the new and old brokers in order to find which configuration parameter caused the high load average.

## Discussion

I/O threads handle requests by picking them up from the request queue for processing. By adding more threads, throughput can be improved, but this decision is influenced by various other factors like the number of CPU cores, the number of disks, and disk bandwidth.

Increasing the number of threads requires careful monitoring of both the NLA (Normalized Load Average) and the CPU `si%` (software interrupt percentage). If the NLA exceeds 1.0 and/or the `%si` is above ±5%, it could indicate that the disk I/O parallelism has been increased excessively.

For clusters with multiple disks per broker, boosting the number of I/O threads generally makes more sense. However, in the context of brokers that have a single disk, we have found that setting the `num.io.threads` configuration to 2 provides an optimal balance. It allows for efficient request processing while maintaining a load average that is smaller than the number of CPU cores (thus the NLA is below 1.0), preventing potential overload.

# Looking at Disk Performance the Whole Time vs. During Peak Time Only

This issue will show why it's sometimes important to look at performance metrics only during peak times, instead of averaging the whole time.

We encountered a production issue in which consumers were lagging during peak times. As the first step, we verified that the load was evenly distributed across all brokers, by checking the average read and write throughput and write IOPS, as can be seen in Figure 7-14.

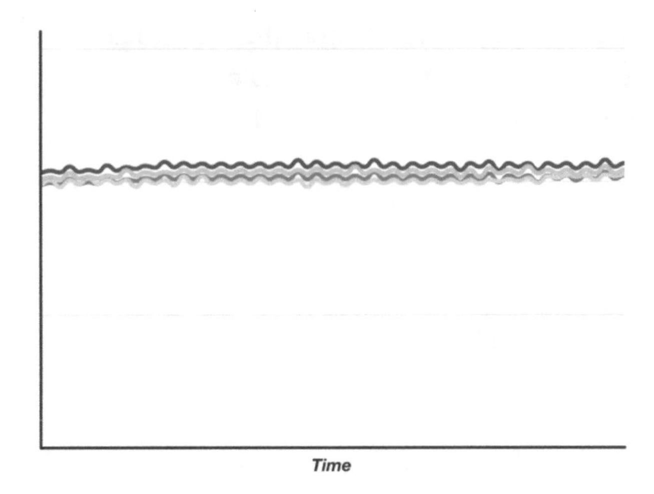


***Figure 7-14.*** *Average sum of writes in MB/sec (wMB/s) in all the disks in a broker, per all the brokers in the cluster*

We also checked the average rMB/s in all the brokers, as shown in Figure 7-15.

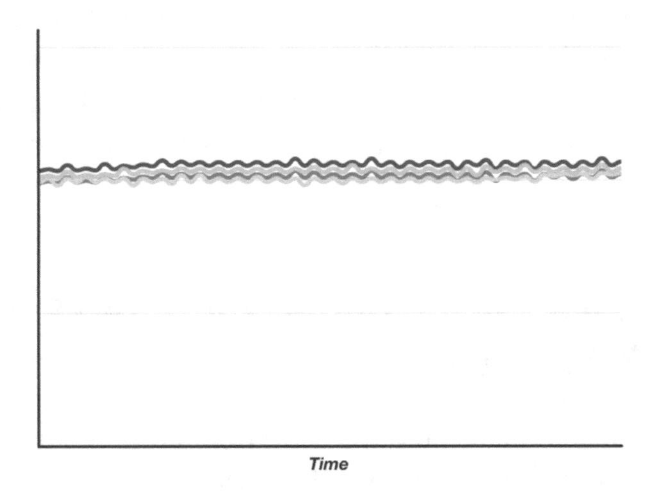


***Figure 7-15.*** *Average sum of reads in MB/sec (rMB/s) in all the disks in a broker, per all the brokers in the cluster*

The average write IOPS/sec in all the brokers is shown in Figure 7-16.

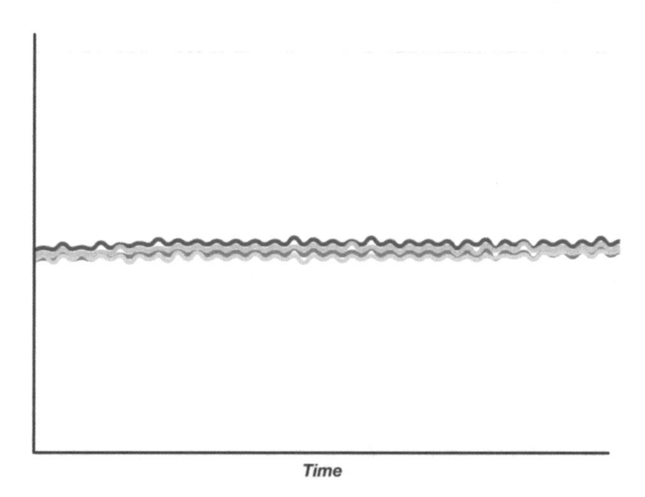


***Figure 7-16.*** *Average sum of write IOPS (W/s) in all the disks in a broker, per all the brokers in the cluster*

The next step was to verify that the even distribution in the writes to the brokers' disks is also reflected in the network throughput, and indeed that all brokers received and sent the same amount of network. This can be seen by the average bytes in all the brokers, as shown in Figure 7-17.

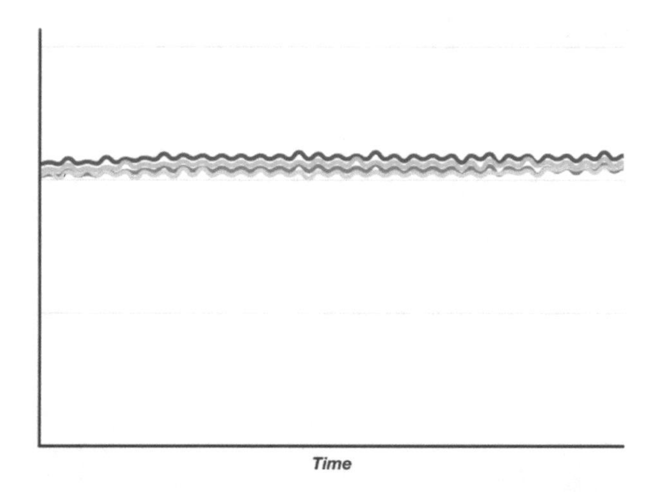


***Figure 7-17.*** *The number of bytes sent to each broker through the network*

Figure 7-18 shows the average bytes in all the brokers.

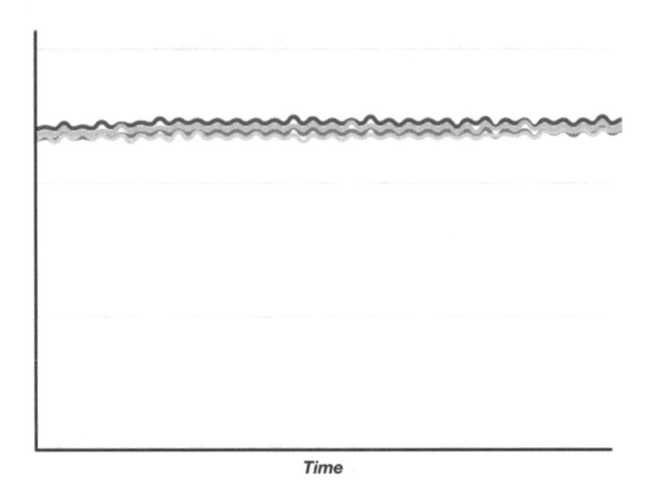


***Figure 7-18.*** *The number of bytes sent from each broker through the network*

After verifying that the disk and network throughput were evenly distributed among the brokers, we proceeded to look at the disk IOPS metric. However, this time, we looked at the max values instead of average values, by checking both the read and write IOPS/sec only during traffic peaks, when consumers lagged. Here, we found an interesting issue in one of the brokers:

Figure 7-19 shows the write IOPS/sec during traffic peaks.

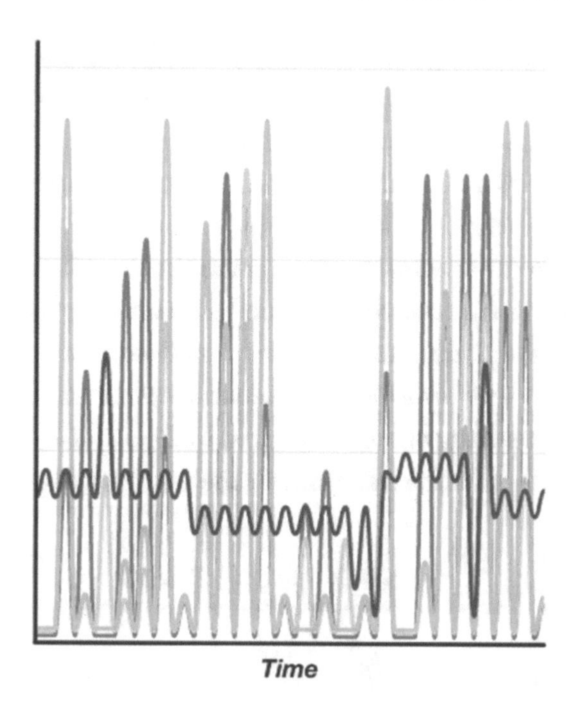


***Figure 7-19.** Write IOPS/sec per each broker during peak traffic time*

Figure 7-20 shows the read IOPS/sec during traffic peaks.

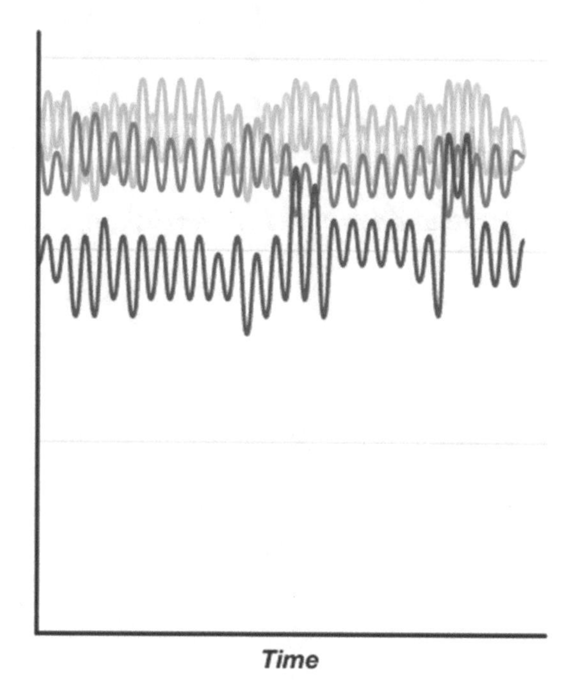


***Figure 7-20.** Read IOPS/sec per each broker during peak traffic time*

As you can see, when looking at the disk IOPS only during peak times, there is a single broker that has significantly lower average read and write throughputs than the rest. It seemed that while other brokers were able to handle bursts, the problematic broker struggled to keep up and managed to perform only a third of the write operations compared to the other brokers.

Then we checked (in the disks of the problematic broker) the time it takes for write and read requests to wait in the queue before being serviced by the disk (using the `r_await` metric for reads and the `w_await` metric for writes, both provided by the `iostat` tool). We noticed that the wait time for read and write operations were ten and thirteen times slower on the problematic broker compared to the other brokers, respectively.

Figures 7-21 and 7-22 show the wait time for the read and write operations in all the brokers.

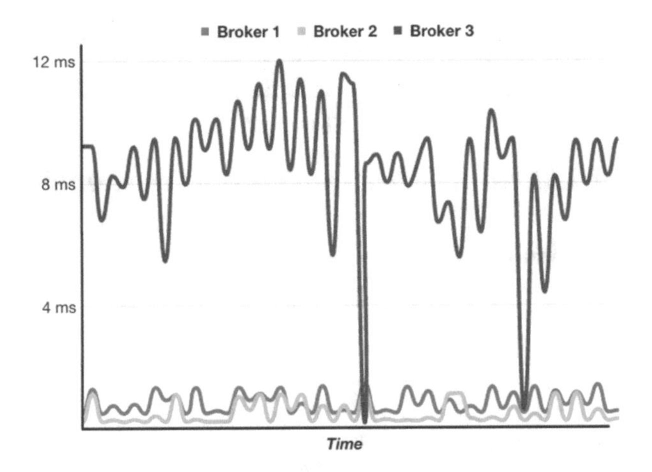


***Figure 7-21.*** *Wait time for read operations on the disks of each broker during peak traffic time*

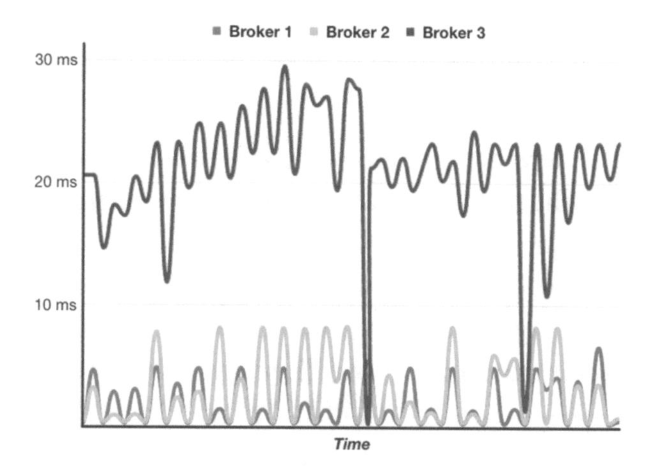


***Figure 7-22.** Wait time for write operations on the disks of each broker during peak traffic time*

After seeing the read and write processing times, we understood that the problematic broker's disk was much slower than the disks of the other brokers.

Another interesting metric was the queue size for produce requests (which is a queue that serves incoming produce requests). This queue was much higher in the problematic broker compared to other brokers, and it was capped, possibly due to the latency on the slow disk.

Figure 7-23 shows the size of the produce request queue in all the brokers.

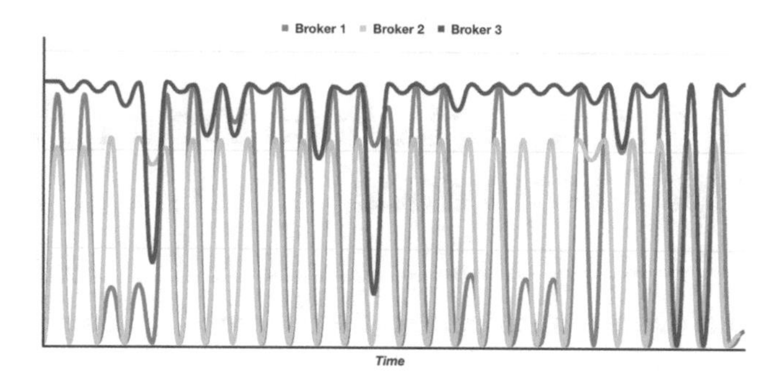


***Figure 7-23.** Size of the produce requests queue in each broker during peak traffic time*

## Discussion

This issue shows the importance of looking at anomalies in a Kafka cluster during peak times, instead of looking at wider time ranges. In this case, we checked the maximum values of the read and write processing times (during the peak traffic time) in order to determine the root cause.

We started with consumers that lagged during peak time, and then the first thing we did was looking at the disk behavior during the whole time, which led to nothing. Only when we looked at the maximum values of disk behavior during the peak time (which has maximum traffic), did we find that the disk of a single broker provided much less write throughput, and that the queue size of writes to that disk was capped. This reduced the throughput of that disk dramatically.

That "slow" disk caused two problems during peak time:

- Due to the high latency on that disk, producers that produced into the partitions that resided on the problematic disk developed an increasing queue of messages in their buffers.
- Due to the low read IOPS, consumers from these partitions started lagging.

Replacing the disk (if the cluster is on-premises) or the broker (if the cluster is on the cloud) will solve this problem.

## The Effect of disk.io Threads on Broker, Producer, and Consumer Performance

This section illustrates an example of how an increase of disk.io threads can impact the cluster. The number of disk.io threads control the number of concurrent disk operations that can be performed on the brokers.

The cluster has six brokers of type `i3en.6xlarge`, each broker has 2x7500 NVMe SSD.

The figures in this section show the impact of increasing the disk.io threads from two to four on several important metrics of the Kafka cluster. These metrics are discussed next.

## Request Queue Size

Request queue size, shown in Figure 7-24, was reduced by a third in all the brokers. This metric shows the number of client requests that wait in the request queue to be processed by the Kafka brokers.

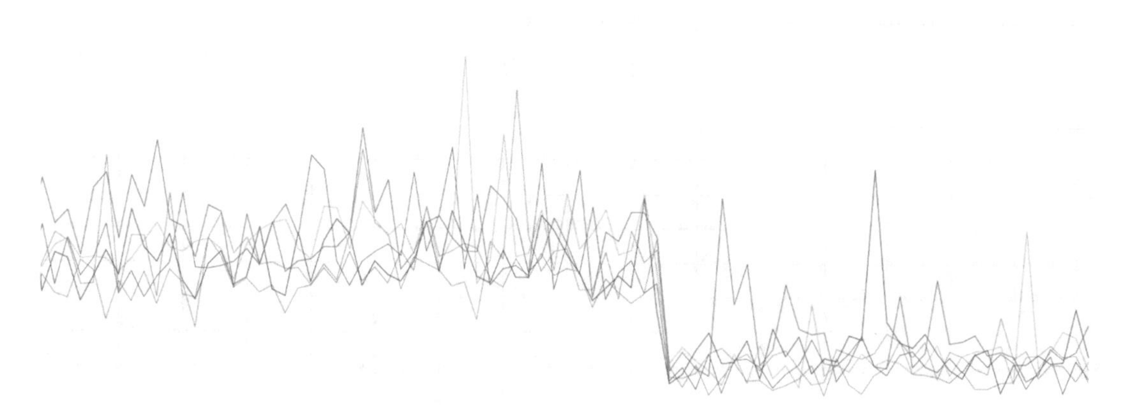


***Figure 7-24.*** *Size of the request queue per broker*

The client requests include produce and consume requests:

- Consumers create fetch request messages and send them to the Kafka broker.
- Producers create request messages and send them to the Kafka broker.

When the broker receives these request messages, they're added to the `RequestQueue` along with any other requests that are waiting to be processed. The broker then processes the requests in the `RequestQueue` in the order they are received

Note that requests from Kafka brokers to fetch data aren't added to the `RequestQueue`, but instead are handled by the Kafka replication protocol, which doesn't use the `RequestQueue` because it operates independently of client requests.

## Produce Latency

Produce latency, shown in Figure 7-25, was decreased by a third in all the brokers. The produce latency metric measures the time it takes for a producer to successfully send a message to the broker. It's measured from the time the message is sent by the producer until it's acknowledged by the broker. This metric includes any network latency, the time the messages waits in the queue, and the processing time of the message.

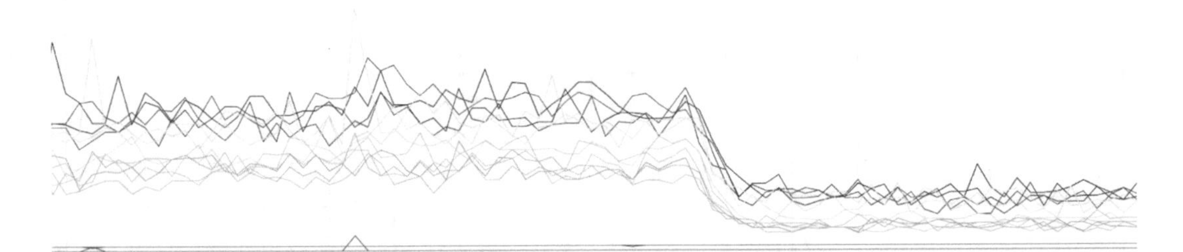


***Figure 7-25.*** *Produce latency (99% percentile) per broker*

## Number of JVM Threads

The number of JVM threads, shown in Figure 7-26, increased two-fold in all the brokers. The increase in the number of disk.io threads reduced the overall latency in the cluster, and it therefore indirectly affected the number of JVM threads used by Kafka, since this allowed Kafka to increase the processing rate of messages and requests.

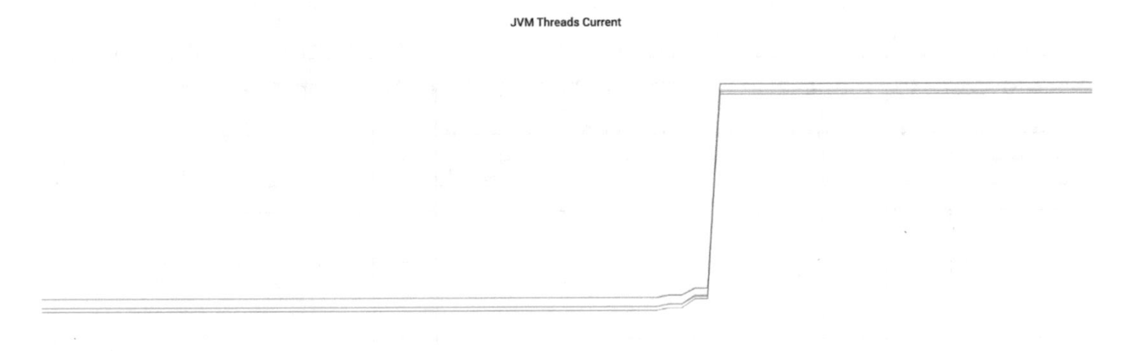


***Figure 7-26.*** *Number of JVM threads per broker*

## Number of Context Switches

The number of context switches, shown in Figure 7-27, doubled. This increase can be attributed to the increase in the number of JVM threads, because as more threads run, there can be more CPU switches between them.

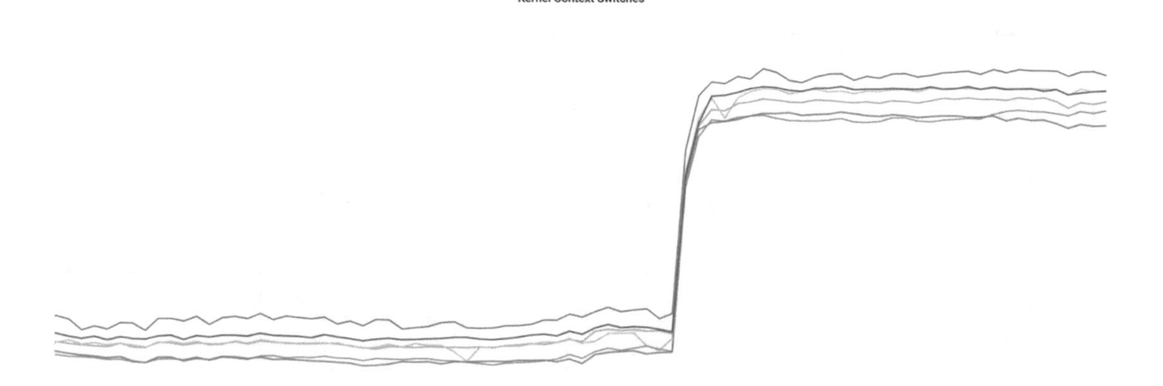


***Figure 7-27.*** *Number of context switches per broker*

## CPU User Time, System Time, and Normalized Load Average

The CPU user time, system time, and normalized load average, shown in Figures 7-28, 7-29, and 7-30, all doubled. This increase can also be attributed to the increase in the number of JVM threads.

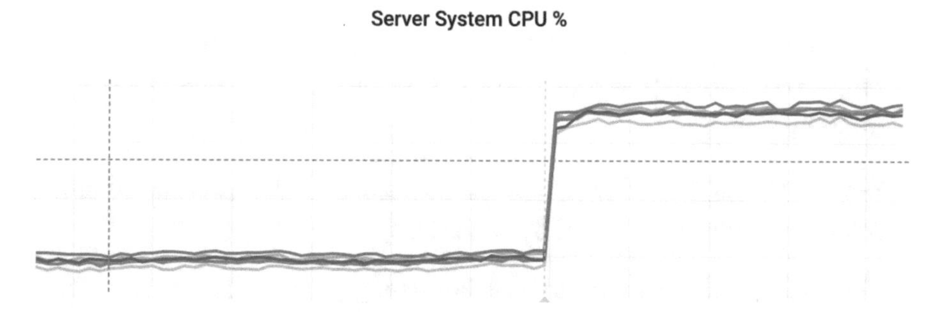


***Figure 7-28.*** *CPU sy% (system time) per broker*

***Figure 7-29.*** *CPU us% (user time) per broker*

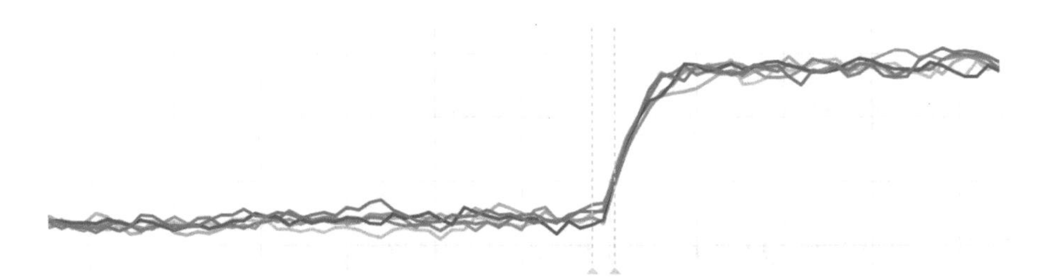


***Figure 7-30.*** *Normalized Load Average (NLA) per broker*

# Discussion

disk.io threads can have a dramatic impact on the overall latency of producers and consumers, and on the Kafka clusters' utilization. In this case, when we switched from a ratio of 1:1 between disk.io threads and the number of disks per broker to a ratio of 2:1, the following occurred:

- The processing rate of client requests (both producer and consumer requests) doubled (the processing rate of fetch requests increased more than that).
- Requests waited much less time in the request queue (the size of the request queue was reduced by two thirds).
- The time it took messages to be acknowledged was reduced by two thirds.
- The CPU utilization of the cluster doubled, and so did the load average of the cluster.

This case shows that tuning the disk.io threads can be beneficial for Kafka clusters when their clients suffer from high latency. However it's important to carefully adjust the number of disk.io threads, since too many threads can cause the brokers to stop functioning. It's also important to pay attention to the overall cluster utilization before trying to increase the diks.io threads, since a cluster that has high CPU utilization will probably not benefit from such an increase, even if its disk utilization is low.

# Summary

This chapter unraveled the intricacies of disk I/O overload and explained how it can impact consumers and producers in Kafka clusters. It began with an introduction to various relevant disk performance metrics, and then delved into how these metrics can be utilized to diagnose issues related to latency in Kafka brokers, consumers, and producers. The analysis provided a deep understanding of how Kafka reads and writes to the disks, shedding light on the factors affecting disk utilization and the symptoms indicating potential disk I/O problems.

The chapter illustrated real-life production issues and demonstrated the practical application of these metrics, such as detecting a faulty broker, and the effect of disk.io threads on the performance of the entire Kafka system. One notable example explored the impact of having too many disk.io threads, revealing the importance of careful adjustment of these threads to prevent overloads.

The chapter concluded by emphasizing the necessity of considering disk performance during peak usage times and the balanced tuning of disk.io threads. These insights should equip you with the knowledge you need to enhance the efficiency and reliability of your Kafka clusters, particularly in detecting and resolving disk-related latency.

The next chapter turns to the choice between RAID10 and JBOD for disk configuration in Kafka production environments. This decision has profound implications for data protection, write performance, storage usage, and disk failure tolerance. Through a thorough comparison of RAID10 and JBOD, we'll examine the tradeoffs and unique benefits of each configuration, including write throughput and disk space considerations. Whether you are prioritizing data security or efficiency in storage and performance, the next chapter will guide you in selecting the option that best suits your Kafka brokers' needs.

CHAPTER 8

# Disk Configuration: RAID 10 vs. JBOD

The decision to choose RAID 10 or JBOD for disk configuration when deploying Kafka clusters is crucial. This chapter thoroughly explores the advantages and disadvantages of both disk configurations in the context of Kafka production environments.

By comparing various aspects of these two configurations—such as disk failures, storage usage, write operation performance, disk failure tolerance, disk health monitoring, and balancing data between disks in the broker—you will gain a comprehensive understanding of the tradeoffs between RAID 10 and JBOD disk configurations.

Ultimately, you will be equipped with the necessary knowledge to make informed decisions as to which disk configuration to opt for in your Kafka production environments.

*JBOD* (*just a bunch of disks*) is a disk configuration whereby the server has internal disks that are controlled individually by the OS, as shown in Figure 8-1. The disks connect to a disk controller on the server, and the disks can be accessed and be seen by the OS.




E. Eldor, *Kafka Troubleshooting in Production*, https://doi.org/10.1007/978-1-4842-9490-1_8

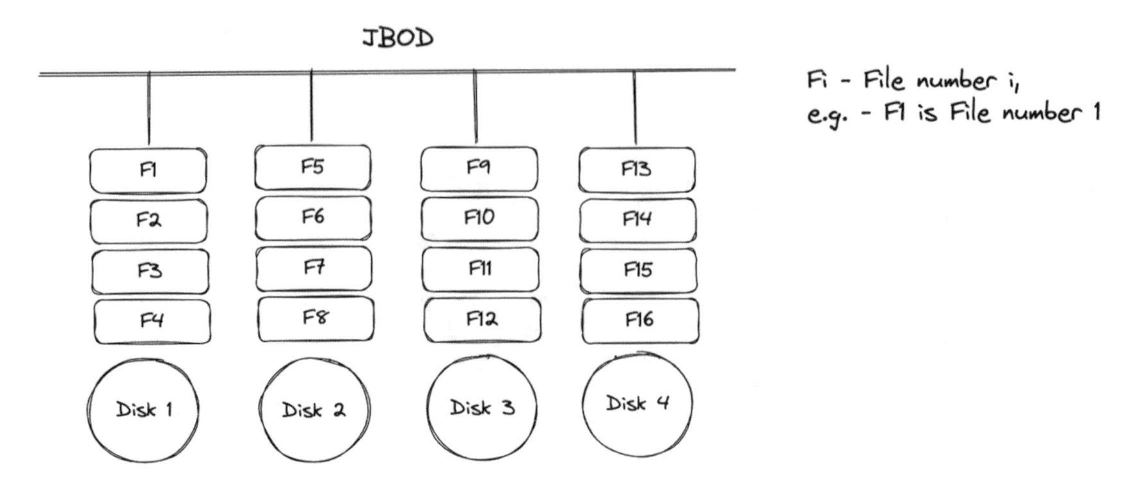


***Figure 8-1.*** *Disks in JBOD configuration*

*RAID* (*Redundant Array of Independent Disks*) presents the disks of the server as a single disk. It can be provided as *hardware* RAID (via a disk controller card) or as a *software* RAID. RAID disks are seen by the OS as a single *virtual* disk.

Before V1.0, Kafka didn't tolerate disk failures, in which case using RAID 10 was almost mandatory since it mitigated the disk failure issue (unless of course two disks in the same mirroring failed). But even after Kafka 1.0, the RAID 10 option is mentioned by Confluent.

# RAID 10 and JBOD Terminology

Because choosing between RAID 10 and JBOD is a tradeoff between several storage-related aspects, this section dives deeper into some of these aspects and presents the pros and cons for each option. In order to do this, the section starts by defining the following terms—RAID 0, RAID 1, RAID 10 (1+0), and JBOD.

## RAID 0 (aka Stripe Set)

RAID 0 distributes the data across the disks, as shown in Figure 8-2. If a file is written to a RAID 0 array that consists of five disks, a fifth of the file will reside on each of the five disks. The advantage of RAID 0 is the speed of writing and reading. In this case, the file can be written and read five times faster than with a single disk.

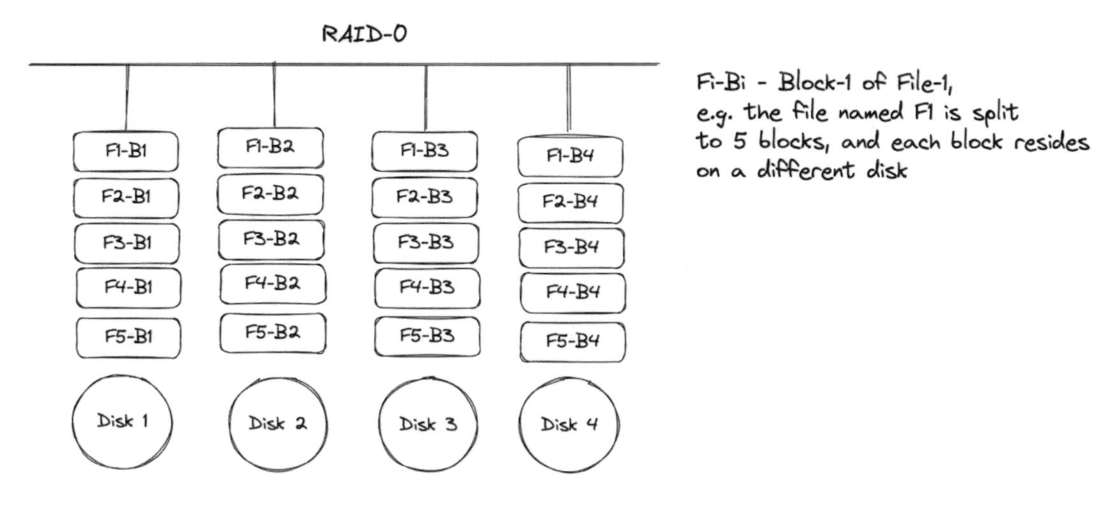


***Figure 8-2.*** *Disks in a RAID 0 configuration*

# RAID 1 (aka Mirror Set)

RAID 1 mirrors the data across an even number of disks, as shown in Figure 8-3. Each pair consists of two disks, and each is an exact duplicate of the other. The write operations are directed to both disks in each pair, which guarantees that the two disks are always in sync. This provides protection against disk failure in case one disk in a pair fails.

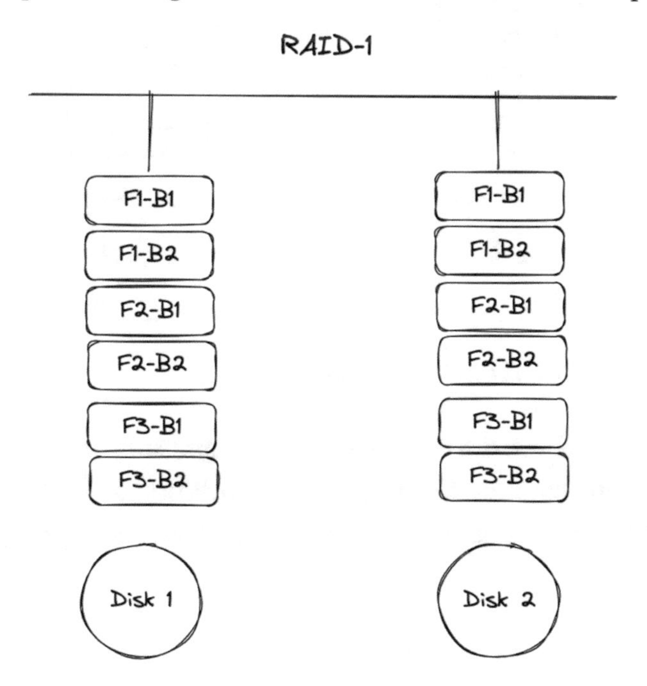


***Figure 8-3.*** *Disks in a RAID 1 configuration*

The storage volume will still be accessible even if half of the disks fail, assuming that each of the failed disks belongs to a different pair. If the two disks of the same pair fail, the volume won't be accessible. The write performance of both RAID 1 and RAID 10 is cut in half (compared to RAID 0) due to the mirroring.

## RAID 1+0 (aka RAID 10)

RAID 1+0 consists of a single stripe set in which all the disks are mirrored pairs, as shown in Figure 8-4.

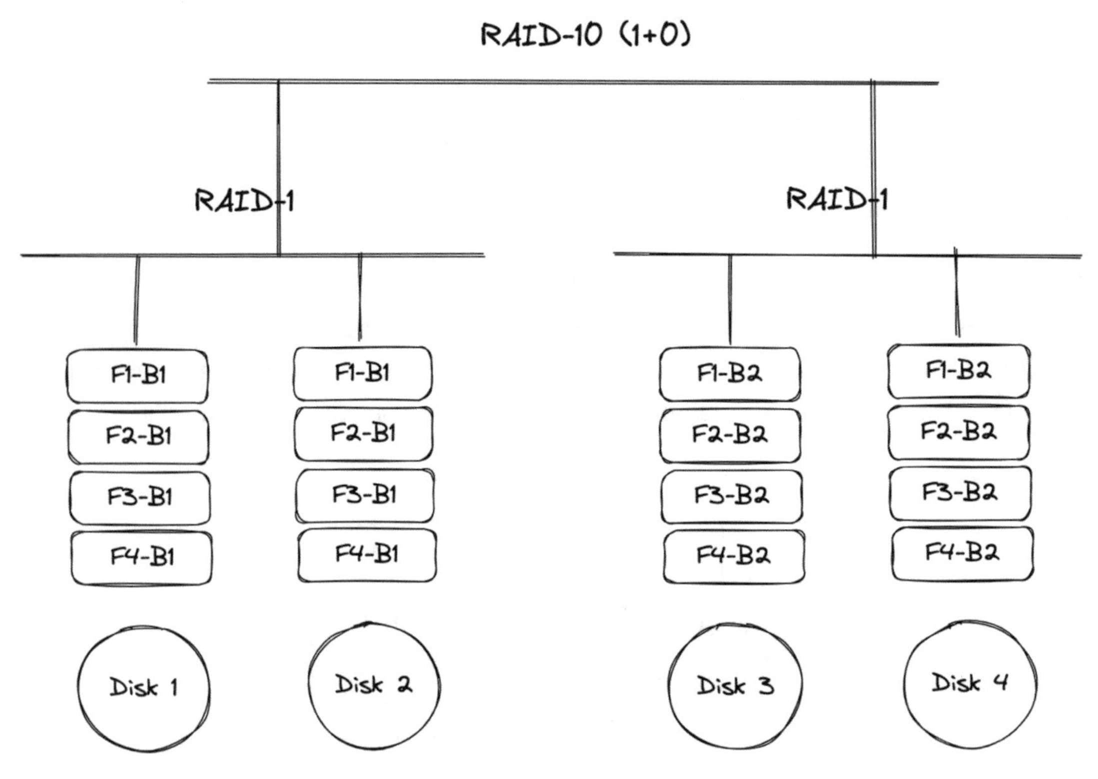


***Figure 8-4.*** *Disks in a RAID 1+0 configuration*

The decision to configure the disks in Kafka brokers as either RAID 10 or JBOD is influenced by two main factors—the level of data protection desired for the information stored in the brokers, and the quantity of storage and disks you are prepared to allocate for that purpose.

# Comparing RAID 10 and JBOD

When it comes to handling disk failure in Kafka brokers, you must consider whether to adopt RAID 10 or JBOD. These configurations determine how the Kafka cluster responds to complete or partial disk malfunctions caused by wear, bad sectors, and other issues. Understanding the protection levels offered by each method is vital. This section examines how JBOD and RAID 10 address data protection, data skew, and storage usage.

## Disk Failure

The more disks are being used, the higher the chances that they'll either partially or completely stop functioning due to some failure, wear out, bad sectors, and so on. Let's consider the level of data protection these two methods provide:

- *JBOD:* There's only one level of protection, which is the replication factor (assuming replication is configured in the Kafka brokers). If a partition leader exists on a disk that fails, one of the followers of that partition will become the leader.
- *RAID 10:* There are two levels of protection—the mirroring (provided by the RAID 1) and the replication within Kafka (assuming replication is configured in the Kafka brokers). In this case, the Kafka cluster can tolerate disk failure better since RAID 10 ensures that the data is replicated across two disks.

## Data Skew

When there is more than a single disk in a Kafka broker, there's a chance for data skew between the disks. The skew can be caused by:

- *JBOD:* If there is more than a single disk per broker, in order to spread the data evenly across the disks per broker, you need to perform this yourself.
- *RAID 10:* Ensures that the data is spread evenly across all disks per broker.

## Storage Use

Let's assume the amount of data written to the brokers is 10GB, and the replication factor for all topics is 2. In that case, the brokers require the following storage:

- *JBOD:* 20GB
- *RAID 10:* 40GB

# Pros and Cons of RAID 10 and JBOD

Choosing between RAID 10 and JBOD isn't a simple decision. To make the right choice, it's essential to weigh the relative pros and cons of these configurations in terms of write operations, storage use, failure tolerance, and on-premises maintenance. This section compares the performance, availability, and manageability of RAID 10 and JBOD, presenting a detailed analysis of each aspect and the conclusions drawn from various tests and real-world experiences.

## Performance of Write Operations

There's a big performance hit for write throughput in RAID 10 compared to JBOD. As a test, I wrote 1M events/sec (1KB in size) into two Kafka clusters—the disks of the first Kafka cluster were configured in RAID 10 and the disks of the second Kafka cluster were configured in JBOD.

The results showed there were 60 percent more writes/sec when using JBOD versus RAID 10. While the Kafka cluster with disks configured in JBOD managed to write 1M events/sec at 60 percent utilization, the same Kafka cluster when configured with RAID 10 managed to write only 400K events/sec at the same disk utilization.

*Conclusion:* This is an advantage of JBOD over RAID 10.

## Storage Usage

When using RAID 10, the disk space being used is twice that compared to JBOD, assuming the replication factor remains the same.

*Conclusion:* This is an advantage of JBOD over RAID 10.

## Disk Failure Tolerance

The strategy to deal with disk failure significantly impacts the resiliency and overall performance of the system. JBOD and RAID 10 have different approaches to handle this:

- *JBOD:* The leaders and followers that reside on a failed disk must be moved to another broker.
- *RAID 10:* The leaders and followers that reside on a failed disk won't be moved because there's mirroring to a second disk, unless it also fails.

The chances of two disks that are part of a pair failing is much smaller than the chance of a single disk failing, which means that there will be less partition movement (whether they're leaders or followers) due to disk failures in clusters configured with RAID 10.

*Conclusion*: This is an advantage of RAID 10 over JBOD.

# Considering the Maintenance Burden of Disk Failure in On-Premises Clusters

This section is relevant only for on-prem clusters, since the maintenance goes on the SREs on the site. There are several aspects to this issue: disk health monitoring, the frequency of replacing disks, and how easy it is to replace the disks. Let's review them one by one.

## Disk Health Monitoring

Monitoring the health of the disks in a Kafka cluster is vital for timely detection and resolution of any underlying issues that might lead to failure. The ability to monitor disks differs between JBOD and RAID 10 configurations.

- *JBOD:* Each disk has its own mount point, and each mount point represents a real disk and not a RAID controller. So it's easier to monitor disks in JBOD configuration.
- *RAID 10:* In the case of hardware RAID, it's not possible to monitor the disks via the OS (e.g. using the SMART tool) because all the disks in the RAID are shown by the OS as one disk, since it's a single mount point. Only by using RAID tools can you check the disk status. (Note: When using software RAID, the disks are visible to the OS.)

## Frequency of Replacing Disks

How often disks need to be replaced in a Kafka cluster depends on the configuration and the monitoring in place. JBOD and RAID 10 configurations have different characteristics that influence the frequency of disk replacement.

- *JBOD:* It's much easier to detect a failed disk, so if there's monitoring on disk failures then every time a disk fails, a new disk will replace the failed disk and mount to the same mount point. With JBOD, the frequency of disk replacement will usually be higher.
- *RAID 10:* It's common that the SRE on site isn't aware that a disk failed (due to the lack of disk failure monitoring). This causes failed disks to not be replaced until two disks of the same mirroring fail (which can cause a production issue).

## Kafka Availability During Disk Replacement

The availability of a Kafka cluster during disk replacement can be affected by the configuration you use:

- *JBOD:* Replacing a disk in a Kafka cluster whose disks are configured in JBOD requires only mounting the new disk without touching the other disks.
- *RAID 10:* Replacing a disk in a RAID controller forces the RAID array to be rebuilt. The rebuilding of the RAID array is so I/O intensive that it effectively disables the server, so this does not provide much real availability improvement compared to JBOD. In fact, I witnessed a higher downtime when replacing disks in RAID than in JBOD.

*Conclusion:* This is an advantage of JBOD over RAID 10.

## Balancing the Data Between the Disks in the Broker

Ensuring that data is evenly spread across the disks in a Kafka broker is crucial for maintaining balanced I/O and optimal performance. Both JBOD and RAID 10 configurations have unique challenges and solutions for balancing data.

## JBOD

It's up to the KafkaOps to make sure data and partitions are spread evenly across the disks per the Kafka broker. If the broker has several data directories, each new partition is placed in the directory with the least number of stored partitions. If the data isn't evenly balanced among the partitions, the data skew can increase.

The weakness of this approach is that it's agnostic to the number of segments per partition. A partition P1 can have more segments than partition P2 due to several reasons. To list a few:

- Assuming partition P1 belongs to topic T1 and partition P2 belongs to topic T2:
    - T1 and T2 have the same incoming traffic but T1 has a higher retention than T2
    - T1 has higher incoming traffic than T2
    - T1 isn't a compressed topic and T2 is compressed
- Assuming both partitions (P1 and P2) belong to the same topic:
    - There's a data skew in the producers writing to both partitions, and P1 receives more traffic than P2.

In order to ensure that data is spread evenly across the disks, you can write a script that reassigns partitions among the brokers based on the number of segments per disk instead of the number of partitions per disk.

## RAID 10

The RAID controller is (theoretically) in charge of balancing the data evenly between the disks. However, according to Confluent's documentation, it doesn't always balance the data evenly. I can't verify Confluent's stand on whether RAID 10 really balances the data evenly across the disks, since I never encountered a case in which the storage of a disk in RAID 10 became full (which is one of the symptoms of data imbalance among disks).

## Managing Disk Health in Kafka Clusters with JBOD Configuration

Utilizing JBOD disks in a Kafka cluster presents unique considerations, particularly regarding disk health monitoring and fault handling. Although this configuration might offer benefits in certain contexts, it also introduces challenges that need careful management.

Kafka's lack of built-in features to recognize problematic disks makes it crucial to understand how a disk's faulty state can impact the cluster. Problems that may arise with disks include transitioning to read-only mode, having bad sectors, failing to mount to specific folders, reaching full storage capacity, and experiencing mechanical issues. Each of these issues requires awareness and appropriate responses to ensure continued cluster operation.

To safeguard the Kafka cluster's reliability when disks in a JBOD configuration are faulty, you must take proactive measures. Since Kafka doesn't inherently support disk health checks, specialized monitoring tools or manual processes may be needed to detect unhealthy disks through kernel messages that alert about disk failures.

One way to respond proactively to such messages is to implement a cron job, which is a scheduled task on UNIX-like operating systems. This cron job can filter these messages hourly, checking for the specific disk failures that need monitoring. If one or more issues are identified, the next step is to disconnect the faulty disk from the JBOD and prevent Kafka from using it. This may also involve procedures for replacing the disk, recovering lost data, or redistributing the load across other disks.

By adopting these measures, you can avoid interruptions in Kafka due to faulty disks and maintain the overall reliability of the Kafka cluster. This may also necessitate engaging with existing tools or third-party solutions for monitoring to ensure that Kafka brokers configured with JBOD function efficiently.

# Summary

This chapter discussed the decision to choose between RAID 10 and JBOD for disk configuration in Kafka production environments.

JBOD is a disk configuration where the server has internal disks that are controlled individually by the OS, whereas RAID presents the disks of the server as a single disk. RAID 10 ensures that the data is replicated across two disks, while JBOD only provides one level of protection via the replication factor (assuming replication is configured in the Kafka brokers).

RAID 10 ensures that the data is spread evenly across all disks per broker, but when using it, there's a big performance hit for write throughput compared to JBOD. Additionally, the disk space used with RAID 10 is double that of JBOD, assuming the replication factor remains the same.

This chapter compared the pros and cons of RAID 10 and JBOD from several angles, including performance of write operations, storage usage, and disk failure tolerance. Ultimately, the decision to opt for one disk configuration over the other depends on the amount of data protection you want for the data stored in the brokers and the amount of storage and disks you are willing to dedicate to that.

The next chapter delves into the essential aspects of monitoring producers in your Kafka cluster. This focused exploration is vital to balancing the relationship between Kafka brokers and producers, which in turn reduces broker load, cuts latency, and enhances throughput. From examining key metrics like network I/O rate and record queue time to investigating the importance of message compression, this next chapter aims to provide valuable insights into achieving optimal Kafka performance.

# CHAPTER 9

# A Deep Dive Into Producer Monitoring

This chapter delves into the art of monitoring producers in your Kafka cluster. Producers play a pivotal role in the Kafka ecosystem, where their efficiency directly influences throughput, latency, and overall resource consumption. A well-tuned producer not only ensures rapid message generation but also sends message in a manner that harmoniously coexists with brokers, avoiding unnecessary disruptions or delays.

This chapter explains a set of metrics that can assist in estimating the stability and performance of Kafka producers, including Network I/O rate, Record Queue Time, Output Bytes, Input Bytes, Average Batch Size, Buffer Available Bytes, and Request Latency.

Additionally, an integral component of producer monitoring lies in the realm of message compression. Proper compression within a Kafka cluster optimizes throughput and minimizes latency while ensuring judicious use of resources. Central to this discussion is the Compression Rate metric, which can be used to estimate the efficiency of the compression strategy.

By the end of this chapter, you'll have a strong grasp of how to monitor the stability and performance of Kafka producers, which will assist you in maintaining smooth data flow in your cluster.

## Producer Metrics

There are several monitoring metrics on the producer side that will assist you in diagnosing issues related to the effect of producers on the Kafka cluster and vice versa.




E. Eldor, *Kafka Troubleshooting in Production*, https://doi.org/10.1007/978-1-4842-9490-1_9

## Network I/O Rate Metric

The Network I/O Rate metric signifies the rate at which data is transferred between the producer and the Kafka brokers. Essentially, it shows how much data is being sent over the network in a given timeframe.

### When Network I/O Rate Is High

When the Network I/O Rate of a Kafka producer becomes notably high, it suggests that the producer is transmitting large volumes of data to the Kafka brokers.

A persistently elevated rate might imply that the brokers are being inundated with data beyond their efficient processing capacity, which can cause an increase in several metrics, such as Consumer Lag, Request Queue, Number of Under-Replicated Partitions, Disk Util% and CPU Util%.

Additionally, it can also cause the network card to get saturated, thereby slowing down processing times and adversely impacting the overall performance of the Kafka cluster.

A High Network I/O Rate metric can be the consequence of several factors. One of the main reasons is the high activity from the producer. If the producer is generating messages at a significantly high rate, this results in a surge in network I/O operations.

Furthermore, the processing speed of the broker might also influence a high network I/O rate, but this influence is a multifaceted issue, shaped by several underlying factors and conditions. If a broker processes incoming messages slowly, the immediate effect might not automatically lead to more frequent network operations. Instead, the producer's response, guided by its specific configuration and error-handling strategies, will define the outcome.

When a broker is slow to acknowledge a message, the producer's reaction can vary. It may try to resend the message, potentially leading to additional network I/O operations. However, this pattern is not consistent across all setups. In some configurations, the producer may simply wait for the acknowledgment (ack) without resending the message, meaning there is no increase in network I/O.

The concept of retries and timeouts in the producer's configuration adds another layer of complexity. More frequent network operations may occur if the producer is set to resend messages after a timeout period while waiting for the ack. Conversely, if the producer just waits without additional actions, the relationship between the broker's processing speed and network operations becomes more nuanced.

Buffer management in the producer also plays a critical role. If the broker's acknowledgment is slow and the producer continues to send messages without receiving acks, the producer's buffer may fill up, potentially leading to data loss if the buffer reaches capacity. This situation requires careful handling and an understanding of the producer's buffering strategy.

To address a high network I/O rate, consider inspecting the producer's configurations and rate of data production. It might be necessary to tune the producer's settings, such as the batch size and linger time. However, note that increasing these settings can create a back pressure on the producers and even cause them to get OOM errors, so these values need to be tuned carefully.

## When Network I/O Rate Is Low

A low Network I/O Rate indicates that data is being sent from the producer to the brokers at a slower pace and that there's more data to be sent. This can be due to various reasons like network issues, producer performance issues, the producer's buffer not being filled quickly enough with data to send to the brokers, or that either the linger time or batch size are poorly configured, which is causing the producers to be idle.

## Importance of the Network I/O Rate Metric

In terms of maintaining high throughput and low latency in the producer, the Network I/O Rate metric acts as a throttle for the rate at which data enters the Kafka cluster from the producer while also serving as an indication of the required throughput. Too much data produced into the brokers can cause an increase in the latency due to processing delays, while too little data can lead to under-utilization of resources and lower throughput.

Therefore, it's important to monitor this metric to ensure optimal use of resources and stability in the Kafka cluster. Figure 9-1 shows a real example of the average network I/O rate among producer pods over a two-week span, which shows a decrease in the network I/O rate after tuning the batch size and the linger time in the producers.

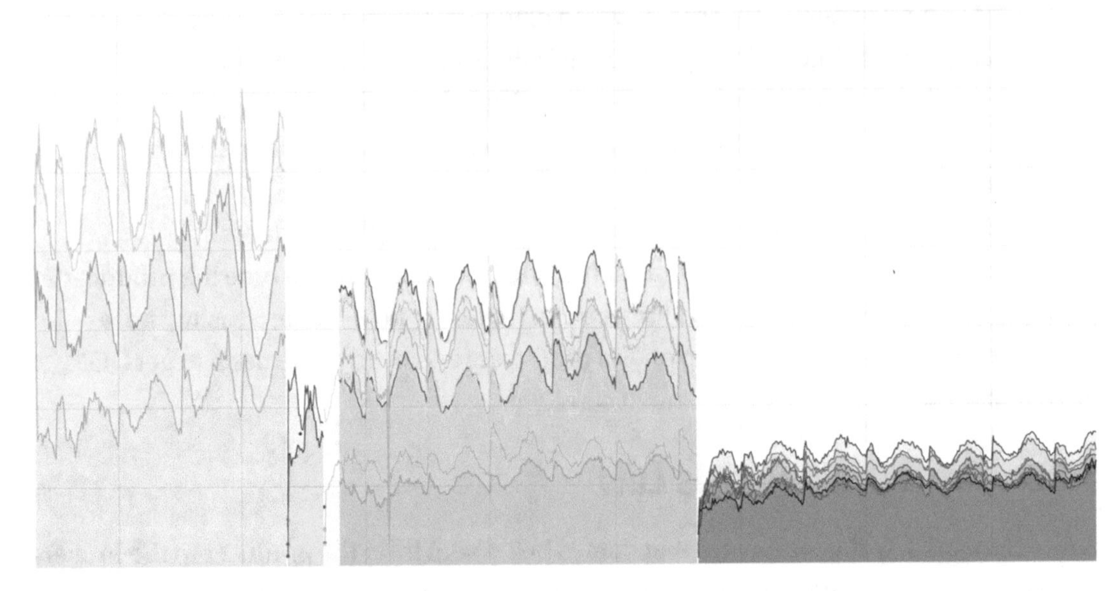

***Figure 9-1.** Average network I/O rate per producer pod. The X axis is the timeline, and the Y axis is the network I/O rate*

## Record Queue Time Metric

The Record Queue Time metric indicates the duration for which a record lingers in the producer's queue before being transmitted to the brokers, as can be seen in Figure 9-2. It's also called record-queue-time-(avg/max) and is provided as part of the JMX metrics.

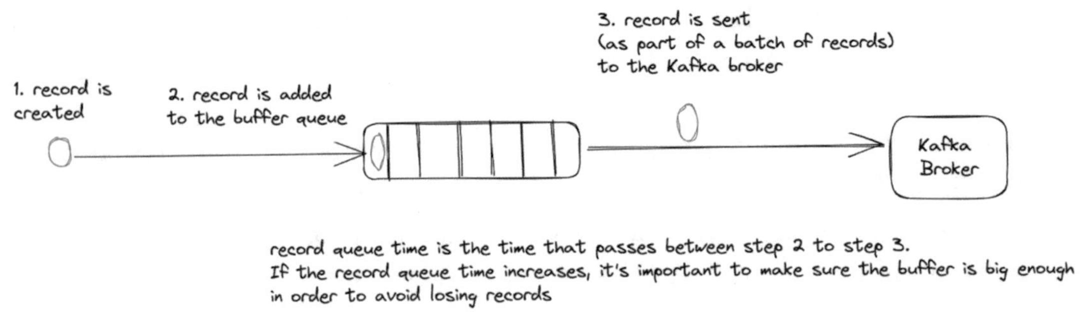


***Figure 9-2.** The record queue time is the time the record waits in the buffer queue*

## When Record Queue Time Is High

When the Record Queue Time increases, it could indicate that records are lingering in the buffer for a prolonged duration.

A high Record Queue Time can be attributed to multiple potential causes.

Network latency might be one of the main factors. If the network between the producer and the broker is slow or congested, records may end up waiting in the buffer for longer periods before being dispatched. Also, the load on the broker can lead to high record queue time. If the broker is swamped with incoming traffic, the processing and dispatch of records to the appropriate partitions might be delayed.

Another contributing factor can be the size of the producer's buffer. If it's too small there can be several implications—messages can't fit into the buffer, which can cause overflow errors; the producer may have to send additional requests to the broker in order to free up space in the buffer; and a reduced throughput due to more requests being sent in smaller batches.

## When Record Queue Time Is Low

A low value signifies that records are swiftly moving out of the producer's buffer and are being sent to the broker, indicating efficient operation.

## Mitigating a High Record Queue Time

If the value is too high, strategies to alleviate this issue include increasing the producer's buffer size, moderating the rate of record generation, or addressing any network bottlenecks.

Increasing the producer's buffer in a Kafka system can have nuanced effects on the record queue time. While increasing the buffer size allows more messages in the queue, aiding in avoiding overflow errors and enhancing batching efficiency, it doesn't necessarily reduce the time that messages remain in the queue. If network bottlenecks or broker overload issues continue, messages might still linger in the buffer, regardless of the buffer's size. If the buffer becomes too large without balancing the record creation rate with the broker's ability to process, it may inadvertently cause increased latency.

Next, moderating the rate of record generation is essential. Effective strategies to balance this rate include implementing a throttling mechanism to control record production, distributing the load across multiple producers to prevent system overwhelm, and employing adaptive algorithms that dynamically respond to various factors like network latency, broker load, and buffer status.

Finally, addressing network bottlenecks forms a critical part of the solution. Regular monitoring and analysis of network performance can lead to actionable insights such as increasing bandwidth, optimizing routing, or scaling network resources.

## Importance of the Record Queue Time Metric

Monitoring the Record Queue Time metric is essential for maintaining a steady flow of data from the producer to the broker. Keeping this metric in check ensures that records are promptly processed, contributing to the stability and performance of your Kafka system. Figure 9-3 shows a real example of the average record queue time among producer pods over a two-week span.

***Figure 9-3.*** *Average record queue time per producer pod. The X axis is the timeline, and the Y axis is the average record queue time*

# Output Bytes Metric

The Output Bytes metric tracks the total number of bytes written to the network by the Kafka producer. It can serve as an indication of the throughput of your data pipeline. This includes the bytes of the messages themselves and the bytes of the protocol overhead.

## When Output Bytes Is High

A high value of the Output Bytes metric indicates a large amount of data being sent from the producer to the Kafka brokers. If this is higher than expected, it might be the result of larger messages (thus fewer messages per batch), more messages, or both. This can lead to network congestion and increased network latency. If the Kafka brokers become overloaded, they will throttle the activity of the producers, preventing them from sending more data than the broker can handle.

## When Output Bytes Is Low

A low value indicates that very little data is being sent from the producer to the Kafka brokers. This reduced data transfer could arise from several reasons. Perhaps there are fewer messages being produced, which might be due to the producer sending messages at a slower rate or because there are fewer active producers.

Slow producers, taking longer to process and send data, can also contribute to fewer bytes being transmitted over time. Issues at the data source, such as disruptions in a logging system or database, might result in a decreased data feed to the Kafka producer.

Additionally, if there are any network constraints or if the Kafka brokers are intentionally slowing the producers to prevent overloading, the volume of messages sent might be limited, leading to a drop in output bytes.

## Mitigating the Output Bytes Value

The Output Bytes metric, whether high or low, isn't inherently indicative of an issue in the Kafka cluster. Instead, it provides a measure of the data throughput into the cluster. Whether these values are problematic depends on the specific expectations and requirements set by administrators and developers.

For some setups, a surge in output bytes might be an expected behavior, while for others, a decline might align with their anticipated data flow. It's crucial for those overseeing the system to determine the appropriateness of these values for their unique use case.

If, after consideration, you decide that the Output Bytes value is indeed problematic, several steps can be taken.

To address high value, you can evaluate the data being transmitted, filtering out non-essential elements or implementing a data sampling strategy. Adjusting the message size or batch size might also be effective. Utilizing data compression before dispatching information to brokers can further optimize the data flow.

Conversely, if the value is lower than desired, it's worth investigating potential causes such as decreased producer rates or smaller message size.

However, if these values align with the system's requirements, no action is necessary.

## Importance of the Output Bytes Metric

Monitoring this metric is crucial for maintaining optimal producer throughput, keeping the Kafka cluster stable, and preventing any potential network or broker overload. Figure 9-4 shows a real example of the outgoing bytes among producer pods over a two-week span.

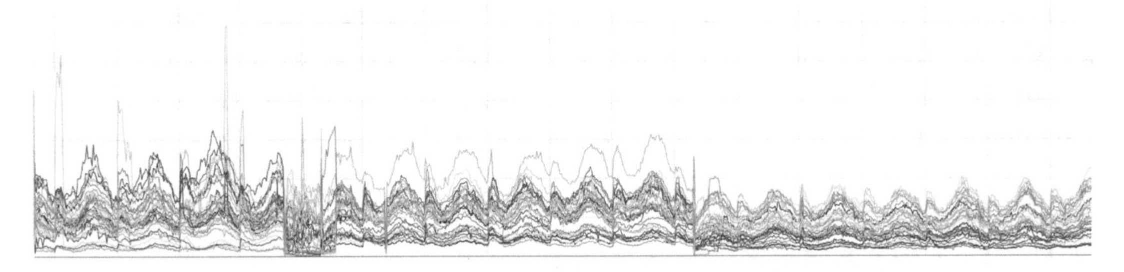

***Figure 9-4.*** *Outgoing bytes per producer pod. The X axis is the timeline, and the Y axis is the outgoing bytes*

# Input Bytes Metric

The Input Bytes metric represents the total number of bytes that a Kafka broker has received from the producers and from the brokers during the replication process. This metric is very similar to the Output Bytes metric (other than the fact that the output bytes include only the bytes sent by the producers without the bytes sent by the brokers during replication) and the same recommendations apply to both metrics.

## The Difference Between Output Bytes and Input Bytes

Before we continue to the next producer metrics to monitor, it's important to understand the difference between output bytes and input bytes.

- The Output Bytes metric (when considered in the scope of producers) refers to the amount of data being sent from the producer to the Kafka brokers.
- The Input Bytes metric (when considered in the scope of brokers) is a measure of the total number of bytes that a Kafka broker has received from producers and from the brokers (during the replication process).

Figure 9-5 shows the difference between these two metrics.

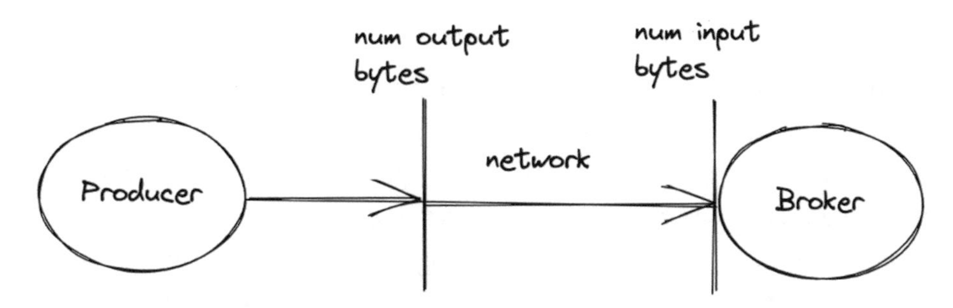


***Figure 9-5.** The difference between num output bytes and num input bytes*

## Average Batch Size Metric

The Average Batch Size metric refers to the mean size of batches in bytes that Kafka producers send to the same topic and partition. By grouping messages into batches, producers can enhance efficiency, as this minimizes the number of network requests, which in turn optimizes throughput and lessens the load on the Kafka brokers. Figure 9-6 shows how messages are turned into batches.

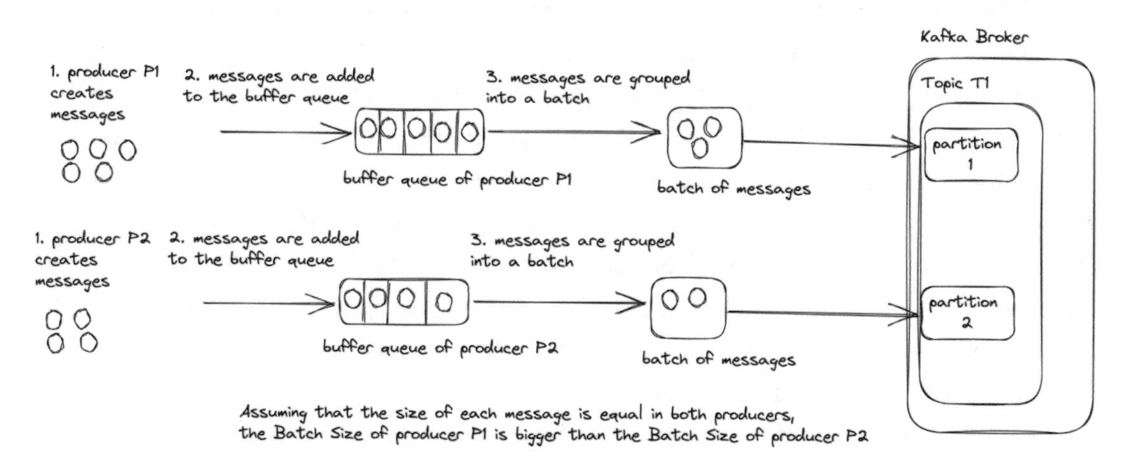


***Figure 9-6.** The flow of a message from its creation by the producer until it reaches its partition in the Kafka broker*

## When Average Batch Size Is High

A higher Average Batch Size metric indicates that the producer is bundling messages into larger batches. This leads to efficient network utilization, as the number of network requests decreases. However, a high Average Batch Size metric can also mean increased latency, as the producer needs to wait longer to aggregate messages into a batch before dispatching them. This is a tradeoff between throughput and latency.

Depending on your specific use case, you might need to adjust this balance. If low latency is a key priority, smaller batches dispatched more frequently can be beneficial. Conversely, if you're prioritizing throughput and network efficiency, larger batches may be more suitable.

## When Average Batch Size Is Low

When Kafka producers have a low average batch size, this denotes that the batches of messages being sent to the same topic and partition are small in size. While small batches can lead to reduced latency, they also make less efficient use of network resources, leading to a larger number of network requests, potentially reducing throughput and increasing the load on the Kafka brokers.

A low average batch size can occur due to several reasons. It might be that the producer is not generating messages fast enough to fill the batches before they are sent. This could be due to insufficient system resources, such as CPU or memory, or it might be that the producer application is not designed to generate messages at a high rate.

Another possible reason is that the producer's configuration settings are leading to smaller batches. For instance, the `batch.size` property in the producer configuration limits the size of the batch, and if this is set too low, it can result in smaller batches. Additionally, the `linger.ms` property controls how long the producer waits to fill the batch before sending. If this is set to a low value, batches may be sent before they are full, resulting in a lower average batch size.

Figures 9-7 and 9-8 show a summary of the potential reasons and negative effects of an average batch size that's either too high or too low.

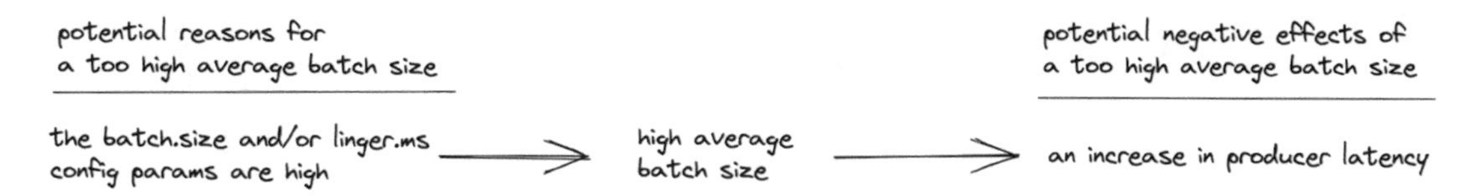


***Figure 9-7.** The potential reasons and negative effects of an average batch size that's too high*

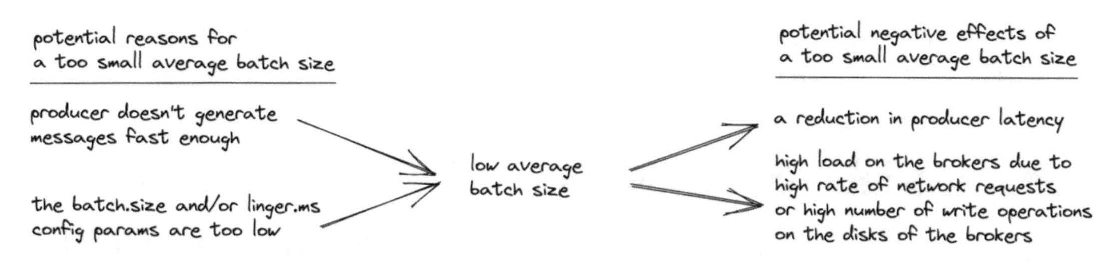


***Figure 9-8.** The potential reasons and negative effects of an average batch size that's too small*

## Mitigating the Average Batch Size Metric

If the Average Batch Size isn't in line with your expectations, you might need to review your Kafka producer configuration. The `batch.size` property sets a byte-size limit for the batch and can be adjusted if necessary. Similarly, the `linger.ms` property can be fine-tuned to impact the average batch size.

## Importance of the Average Batch Size Metric

Monitoring the Average Batch Size metric can provide insights into the efficiency of your Kafka producer batching. It helps maintain a balance between network efficiency and latency, which is crucial for the overall performance of your Kafka setup. Figure 9-9 shows a real example of the average batch size among producer pods over a two-week span, during which several tweaks to the batch size and linger time were performed in order to increase the batch size.

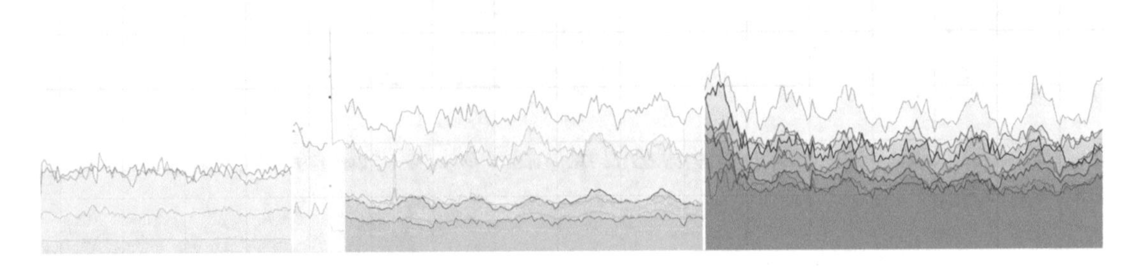

***Figure 9-9.*** *Average batch size per producer pod. The X axis is the timeline, and the Y axis is the average batch size*

# Buffer Available Bytes Metric

Buffer available bytes is the amount of free space in the producer's buffer that's available for new records.

## When Buffer Available Bytes Is High

A high value suggests the producer can send data to the broker as quickly as it's being produced, resulting in an empty buffer ready to accept new records. This is typically a healthy sign, but it can also mean the cluster is over-provisioned.

## When Buffer Available Bytes Is Low

A significantly low value for the Buffer Available Bytes metric, especially if it approaches zero, signifies that the buffer is nearly full. This could suggest that the producer cannot dispatch data as rapidly as it's being produced. The likely culprits in such a scenario could be network or broker issues or producer performance.

Several factors can contribute to a smaller Buffer Available Bytes metric. The rate at which the producer generates messages can be a major reason. If messages are being generated at an exceptionally high rate, it can drastically shrink the available buffer size. Additionally, the size of the messages is also a significant factor.

Oversized messages can fill up the available buffer space rapidly. A slow network might also be causing this issue. If the network is slow, it can delay the dispatch of messages from the producer to the broker, leading to an accumulation of messages in the buffer.

Another reason can be a slowdown in the broker that processes the incoming messages, which can result in a backlog in the producer's buffer.

### Mitigating the Buffer Available Bytes Metric

If the value is too low, consider increasing the buffer memory size in the producer configuration, or check for issues that are slowing down the producer's send rate.

### Importance of the Buffer Available Bytes Metric

Monitoring the Buffer Available Bytes metric helps detect producer bottlenecks and ensures the smooth flow of data from the producer to the broker.

## Request Latency (Avg/Max) Metrics

The Request Latency (Avg/Max) metrics represent the average and maximum latencies for requests made to the broker.

### When Request Latency Is High

High request latency could be a sign of network complications or an overloaded broker, both of which can result in extended record send times and reduced throughput. High network latency could be a major reason, as it can increase the time taken for the request to travel from the producer to the broker and back. Furthermore, if the broker is overloaded with messages, it might take longer to process each request.

High throughput could also be a contributing factor. If the producer is pushing vast amounts of data, it can increase the request latency, since each request contains more data to process and the brokers can't handle this amount of requests. Lastly, resource contention in a multi-tenant environment can be another reason. If other processes are consuming a substantial part of the resources (CPU, memory, disk I/O), this could lead to increased request latency.

### When Request Latency Is Low

A low value suggests the producer-broker communication is happening swiftly, indicating a healthy system.

### Mitigating the Request Latency Metric

If the value is too high, check the network and the broker for potential issues.

### Importance of the Request Latency Metric

Monitoring the Request Latency Avg/Max metrics is crucial to ensuring timely data delivery and high system performance.

## Understanding the Impact of Multiple Producers and Consumers on the Kafka Cluster

In our exploration of producer metrics and Kafka cluster performance, it's crucial to understand that a Kafka cluster doesn't operate in isolation. The performance of an individual producer cannot be fully understood without considering the broader context of the ecosystem in which it operates.

Indeed, a Kafka cluster typically consists of multiple producers and consumers working in tandem. Each of these entities sends, receives, and processes data simultaneously, adding complexity to the dynamics of the cluster. This multi-actor environment can result in a scenario where high metric values observed for one producer might not necessarily stem from issues with that producer itself.

Let's illustrate this with an example: Suppose your producer is experiencing elevated network I/O rate or record queue time values. While it's tempting to attribute these anomalies directly to your producer's configuration or workload, they might, in reality, be a consequence of a high traffic situation caused by other noisy producers or consumers in the cluster.

In this scenario, the busy and noisy neighbors are effectively saturating the brokers' capacity to handle incoming traffic or requests, thus impacting all producers' ability to efficiently dispatch their data. As a result, metrics for your producer could seem problematic, even if the producer is functions as expected.

The takeaway here is that effective Kafka cluster monitoring requires a holistic approach. While focusing on individual producer metrics is undoubtedly important, considering the wider context of the entire cluster—including the activities of other producers and consumers—is equally critical.

## Compression Rate: A Special Kind of Producer Metric

Compression rate gets its own section due to the importance of compression of messages in a Kafka cluster. This section delves into who can and should perform compression in a Kafka cluster.

Compression can be performed either by the producers or by the brokers. However, the general best practice is for the producers to compress the messages before sending them, with the brokers simply passing on the compressed messages.

The reasoning behind this approach is efficiency and resource utilization. If producers compress the messages before sending, it reduces the amount of network bandwidth needed to transmit the messages. This can lead to better throughput and lower latency, which is particularly beneficial in high-volume environments.

Furthermore, performing compression at the producer level also reduces the load on the Kafka brokers. Instead of having to handle compression on top of their other responsibilities (like distributing messages to consumers), the brokers can focus more on these tasks, leading to better overall performance.

It's worth noting that not all types of data compress well, so the effectiveness of compression can vary depending on the nature of the data you're working with. Therefore, it's best that you understand the characteristics of your data and perform testing to ensure that the benefits of compression outweigh its costs.

Remember that the choice of compression type (e.g., Gzip, Snappy, LZ4, or Zstd) can also impact both the compression rate and the CPU load of the producers and brokers. Different compression algorithms offer different tradeoffs between CPU usage and compression rate, so you should choose the one that best fits your specific use case and resources.

Finally, it's important to monitor the Compression Rate metric, as it provides insight into the effectiveness of your compression strategy. A high Compression Rate means your compression strategy is working well, whereas a low rate could indicate that the compression algorithm isn't effective for your particular data.

## Configuring Compression on the Producer and Broker Levels

Configuring compression at the producer and the broker level is a common mistake, often rooted in a misunderstanding of how Kafka's compression works. This approach can lead to unnecessary overhead and, paradoxically, may even reduce the performance benefits provided by compression.

Kafka uses a "write once, read many" model. When a message is produced, it's written to a Kafka topic where it can be read by many consumers. This means that if a message is compressed before it is written to a topic, every consumer can benefit from that compression without needing to decompress and recompress the message. Thus, implementing compression at the producer level can maximize efficiency.

When compression is configured on both producers and brokers, the messages will be compressed by the producer, decompressed by the broker, and then compressed again before being sent to the consumers. This is a wasteful process that consumes unnecessary CPU resources for the additional decompression and recompression steps.

Therefore, it's generally best practice to enable compression only at the producer level, while brokers are simply tasked with forwarding the already compressed messages to consumers. This approach optimizes resource usage and maintains the integrity of message order.

## Compression Rate

Compression rate is the ratio of compressed bytes to uncompressed bytes, measuring the effectiveness of the compression performed by the producer. This section assumes the compression is performed by the producers and not by the brokers. Figure 9-10 shows the flow of the compression process.

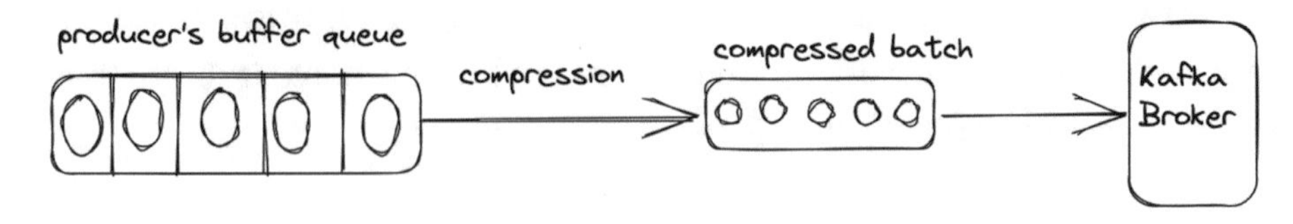


***Figure 9-10.*** *The messages in the batch are compressed and then the batch is sent to the Kafka broker*

## When Compression Rate Is High

A high value indicates that the producer is efficiently compressing data, which can reduce the amount of network bandwidth needed and potentially save storage on the Kafka broker. However, higher compression also means increased CPU usage for compressing (on the producer side) and decompressing messages (on the consumer side).

## When Compression Rate Is Low

A low value indicates that the producer is not compressing message data very efficiently before sending it to the Kafka brokers. Compression can help reduce the size of the data being sent over the network, improving network utilization, and reducing the load on the Kafka brokers. However, it also requires additional CPU resources to compress and decompress the data.

A low compression rate might be due to the nature of the message data being sent by the producer. Some types of data are more easily compressible than others. For example, text data is typically highly compressible, while binary data or already compressed data may not be.

Alternatively, it could be due to the `compression.type` configuration setting in the producer. This setting determines the type of compression algorithm used by the producer, and different algorithms have different compression efficiencies. For instance, Gzip generally achieves a higher compression rate but uses more CPU resources, while Snappy and lz4 have lower compression rates but are faster and use less CPU.

In some cases, a low compression rate might be desirable if the goal is to minimize CPU usage at the expense of network utilization. However, in environments where network bandwidth is a limiting factor, it may be beneficial to adjust the producer configuration or message data to achieve a higher compression rate.

## Mitigating Compression Rate

If the Compression Rate value is too low, consider changing the compression type or adjusting the compression settings. If the value is too high and CPU usage is a concern, consider using a less CPU-intensive compression type.

## Importance of the Compression Rate Metric

Monitoring the Compression Rate metric can help balance the tradeoffs among network usage, storage needs, and CPU usage.

# Summary

This chapter provided a comprehensive guide to monitoring producers in a Kafka cluster, since effectively monitoring the producers can reduce the load on the brokers, decrease latency in producers, and increase overall throughput.

The chapter discussed several key producer metrics, including Network I/O Rate, Record Queue Time, Output Bytes, and Average Batch Size.

Additionally, the chapter devoted a special section to the Compression Rate metric, highlighting the importance and effectiveness of message compression in a Kafka cluster.

The upcoming chapter turns your attention to consumer monitoring in the Kafka cluster. It dives deeper into the metrics and behaviors essential to consumers, ensuring smooth message flow and efficient processing. It explores essential consumer metrics, including Consumer Lag, Fetch Request Rate, and Bytes Consumed Rate, which together provide insights into the Kafka consumers' functionality. It also delves into the relationships between consumer metrics and other Kafka components, as well as discusses the complexities of data skew and consumer lag. To bring these concepts to life, the chapter includes a case study highlighting how to pinpoint and address broker overload using metric correlations.

# CHAPTER 10

# A Deep Dive Into Consumer Monitoring

This chapter thoroughly examines the important elements for monitoring Kafka consumers in Kafka clusters. It discusses specific metrics and their interplay, because understanding these dynamics is key to maintaining a well-balanced relationship between Kafka brokers and consumers. By focusing the efforts on reducing broker load, decreasing consumer latency, and enhancing overall throughput, you'll be on the path toward increasing the performance and stability of your Kafka cluster.

Ensuring consumers efficiently receive and process messages is paramount to preventing data traffic congestion and maintaining smooth data flow throughout the cluster. Through vigilant monitoring and timely intervention, you can swiftly identify and address potential issues, enhancing the overall reliability of your Kafka cluster.

This chapter first focuses on the consumer metrics that need careful consideration. The chapter then discusses the relationship between data skew in partitions and consumer lag. This is a common scenario in Kafka clusters, whereby you might encounter different combinations of consumer lag and data skew across the topic's partitions, which are being consumed by the consumers. Understanding these situations and knowing how to manage them can substantially boost the health of your data streaming pipeline.

Finally, the chapter presents a practical example of correlating consumer, producer, and broker metrics. I recount my experience with a Kafka cluster where one of the brokers was receiving significantly more write operations than the others. By correlating a set of specific metrics, I was able to pinpoint this issue and tackle it effectively.




E. Eldor, *Kafka Troubleshooting in Production*, https://doi.org/10.1007/978-1-4842-9490-1_10

# Consumer Metrics

This section reviews several Kafka consumer metrics that provide essential information about message consumption rates, fetch details, latency, and various other aspects. These metrics are instrumental in ensuring the smooth functioning of consumers, and by extension, the producers and brokers in a Kafka setup. Furthermore, they provide insights into the system's performance, efficiency, and reliability, enabling you to optimize resource usage, improve system responsiveness, and maintain high-quality data flow in your Kafka pipeline.

## Consumer Lag Metrics

Consumer Lag metrics represent the difference between the last produced message and the last consumed message by a specific consumer, as shown in Figure 10-1.

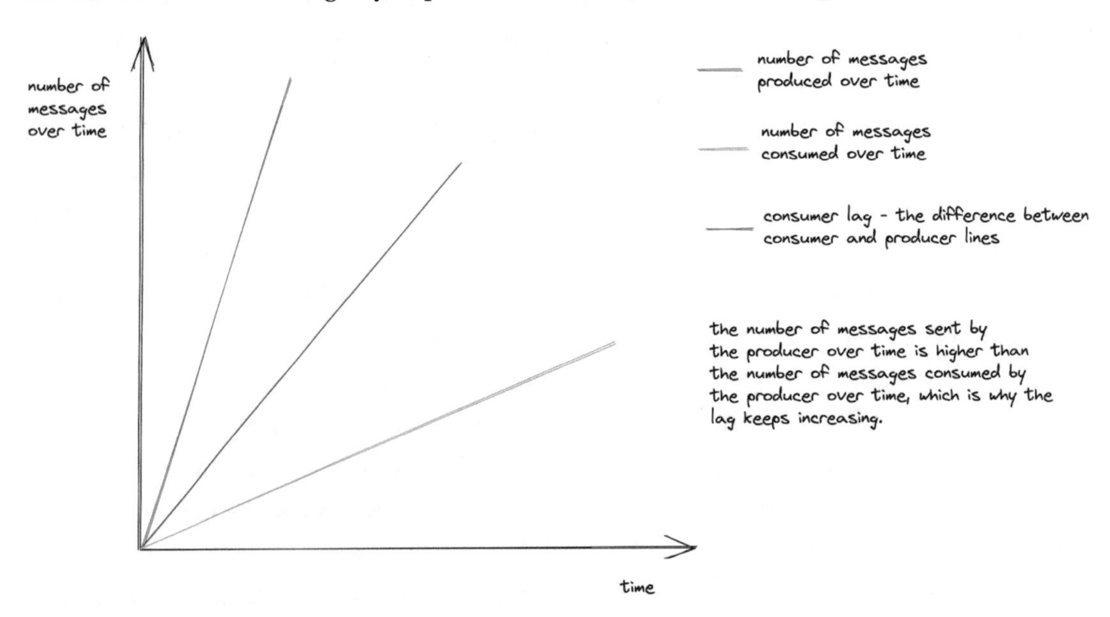


***Figure 10-1.*** *As long as the number of produced messages is higher than the number of consumed messages, the consumer lag increases*

## When the Consumer Lag Metric Is High

A high consumer lag in Kafka is specific to each consumer and the topic it subscribes to. It indicates that a particular consumer is not processing messages as quickly as they are being added to the topic, leading to a growing backlog of unread messages.

This metric specifically measures the difference between the position of the latest message added to the topic and the last message read by the consumer. A growing gap implies that the consumer's processing rate is lagging behind the message production rate for that topic.

There could be several reasons for a high consumer lag. It might be from slow consumer processing, due to factors like inefficient code, insufficient resources, or a misconfigured consumer setup.

Extended processing time in the consumer can also contribute. Persistently high consumer lag can hinder the timeliness and reliability of your Kafka-based systems, potentially causing outdated data processing or delays that might impact the real-time responsiveness of your application.

## When the Consumer Lag Metric Is Low

A low value suggests the consumers are successfully processing messages at or near real-time.

## Mitigating the Consumer Lag Metric

If the value is high, consider scaling up the consumers or optimizing their processing logic.

## Importance of the Consumer Lag Metric

Monitoring the Consumer Lag metric is essential to ensure real-time processing and identify potential bottlenecks.

Understanding how to monitor consumer lag depends on whether you're dealing with low-level consumers or high-level consumers.

- Low-level consumers in Kafka manually manage the offsets, allowing them to decide when to move on to the next message. They also have the flexibility to replay or skip messages if desired. Monitoring consumer lag of low-level consumers can be challenging because it necessitates manually keeping track of the offsets. The lag needs to be calculated by finding the difference between the highest message offset in the topic and the last offset read by the consumer for that specific topic partition.

- High-level consumers utilize Kafka's built-in consumer groups to handle offsets. Kafka inherently tracks the highest offset the consumer group has processed, facilitating the computation of consumer lag by contrasting this value with the most recent message offset in the topic.

To sum it up, while monitoring consumer lag on high-level consumers is relatively straightforward due to built-in tracking, doing the same on low-level consumers demands additional manual oversight and computations.

## Fetch Request Rate Metric

The Fetch Request Rate metric indicates the number of fetch requests a consumer sends to a broker per second

### When the Fetch Request Rate Is High

A high fetch request rate suggests that the consumer is frequently requesting data from the broker.

- From the consumer's perspective, a high fetch request rate might lead to higher network traffic and increased load on the consumer as it has to process the fetched data.
- From the broker's perspective, it has to handle a large number of incoming fetch requests, which can potentially overload the broker and increase response time.

Potential causes for a high fetch request rate can be a high message production rate or a low `fetch.min.bytes` configuration on the consumer, which leads to more frequent requests, because the consumer fetches data as soon as the specified amount of data is available.

## When the Fetch Request Rate Is Low

A low value may indicate under-utilization of the consumer's fetch capacity or possible network problems.

## Mitigating the Fetch Request Rate

The `fetch.max.bytes`, `fetch.min.bytes`, and `fetch.max.wait.ms` configuration parameters play a critical role in optimizing the performance of consumers, and they directly impact the fetch requests made by the consumers.

The `fetch.max.bytes` parameter sets the maximum amount of data the server should return on a fetch request. Increasing the value of `fetch.max.bytes` will allow a consumer to pull more data in each request, reducing the overall number of requests, which could be beneficial when network overhead is high. However, it will also increase memory use, as more data will be held in memory.

The `fetch.min.bytes` parameter sets the minimum amount of data the server should return on a fetch request. If not enough data is available to meet this minimum, the request will wait until sufficient data is available. This can reduce the number of fetch requests when the data production rate is low, effectively saving CPU and network resources. However, setting `fetch.min.bytes` too high might lead to delays in message delivery when data production rate is not consistently high.

The `fetch.max.wait.ms` parameter determines the maximum amount of time, in milliseconds, that the broker will wait before answering a fetch request when there isn't enough data to satisfy the `fetch.min.bytes` value. Essentially, it balances between latency and throughput. A shorter wait time can lead to quicker data delivery, but may result in more frequent, smaller fetches. Conversely, a longer wait time allows for larger batches of data to be fetched, which can be more efficient but might introduce a delay in data reception.

If you have a consumer in a Kafka cluster that's making too many fetch requests, which is shown by a high fetch request rate, there are a few changes you can make.

- First, you can increase `fetch.max.bytes`, which lets each fetch request get more data at once. This means fewer fetch requests are needed, which can lower the fetch request rate.

- Second, you can increase fetch.min.bytes if your data production rate changes a lot. This makes each fetch request wait until there's enough data to send, which means you'll have fewer fetch requests when your data production is low.
- Third, think about how big the messages you're producing are. If they're small, you're going to have a lot of fetch requests. If your situation allows it, you might want to think about grouping smaller messages together to cut down on the number of fetch requests you need to make.

Figure 10-2 shows the effect of low and high values of fetch.min.bytes and fetch.max.bytes in the consumer.

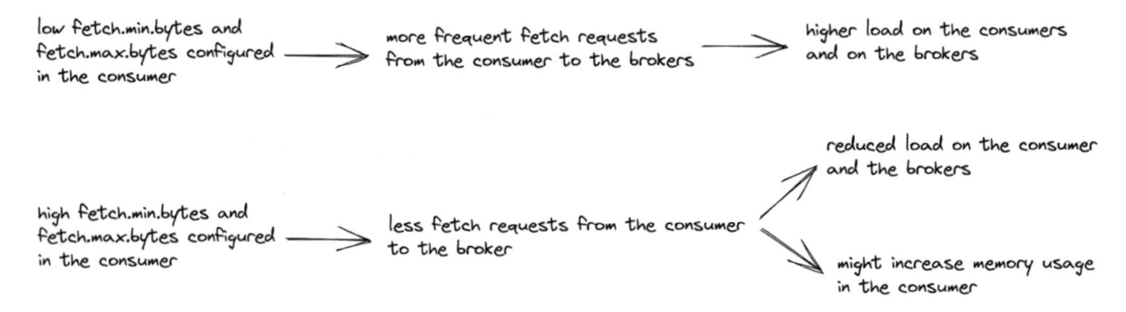


***Figure 10-2.*** *How low or high values of fetch.min.bytes and fetch.max.bytes affect the load and memory use of the consumer*

If your consumer makes frequent fetch requests but retrieves smaller amounts of data, or if you're looking to manage the tradeoff between data delivery latency and fetch efficiency, adjusting fetch.max.wait.ms can be beneficial.

By increasing the wait time, the consumer allows for a potentially larger amount of data to accumulate before fetching, which could reduce the fetch request rate and improve overall fetch efficiency. However, you should be mindful, as setting this too high can introduce noticeable delays in data delivery, especially when the data production rate isn't consistently high.

## Importance of the Fetch Request Rate

Monitoring the fetch request rate helps you optimize consumer performance and resource utilization.

## Fetch Request Size (Avg/Max) Metrics

The Fetch Request Size Avg/Max metrics measure the size of the fetch requests made by a consumer.

### When the Fetch Request Size Is High

High values for these metrics suggest that the consumer is fetching large amounts of data with each request.

For the consumer, handling larger fetch requests means it requires more memory, as it must hold onto the incoming data for processing. If the consumer isn't set up to manage such large volumes of data, several problems might arise.

Slower processing time can result from the challenges of managing and working on bigger datasets, especially when the consumer has other tasks to perform at the same time.

If the fetched data size exceeds the consumer's memory limits, especially its heap memory, it might encounter out-of-memory errors.

Further, there's an increased chance of higher latency. This can be due to the longer time needed to process substantial data chunks, delays in resource allocation for efficient data management, and increased garbage collection activity, as the consumer uses more memory.

Fetching larger data batches is often more efficient for the broker as it minimizes the strain of handling many smaller requests. Given the broker's capacity, it can accommodate these heftier fetch requests without additional stress or compromising response time.

### When the Fetch Request Size Is Low

A low value suggests that the consumer is fetching smaller amounts of data more frequently. This can lead to inefficient use of network resources and can increase CPU cycles due to the constant processing of these smaller requests. Moreover, the system might experience more context switches as it has to frequently handle these requests, which can add overhead and reduce overall efficiency.

## Mitigating the Fetch Request Size

The `fetch.max.bytes` and `max.partition.fetch.bytes` configuration parameters directly influence the fetch requests made by the consumers.

The `fetch.max.bytes` parameter is discussed in detail in the section about the Fetch Request Rate metric. It determines the maximum amount of data the consumer fetches in one request from the broker. Increasing this value allows a consumer to pull more data in each request, potentially reducing the number of fetch requests. This can be especially beneficial when network overhead is a concern. However, it's crucial to balance this against consumer memory capacity and ensure that the consumer can handle larger batches of fetched data.

The `max.partition.fetch.bytes` parameter sets the maximum amount of data per partition that the broker returns. It should always be larger than the maximum message size the server allows, or the consumer may not be able to consume messages. When dealing with a high fetch request size, increasing this value allows consumers to pull more data per partition in each fetch request. However, this also means that a larger batch of data is kept in memory before being processed, which may impact consumer memory use.

If your consumer retrieves significant data volumes with each fetch request, as evidenced by a high fetch request size, these parameters can be tuned in order to reduce the fetch request size. However, it's important to emphasize that fetch request size doesn't have a universal optimum. The adjustments are about tailoring the consumer's behavior to specific needs rather than rectifying a fundamental issue.

- First, if network resources and consumer memory usage are not an issue, increasing `fetch.max.bytes` can potentially decrease the number of fetch requests required to consume the same amount of data.
- Second, you could also consider increasing `max.partition.fetch.bytes` if the data produced in each partition is large. This allows each fetch request to pull more data per partition, reducing the number of fetch requests required to consume the data across all partitions. However, always ensure your consumer has sufficient memory to hold the fetched data.

## Importance of the Fetch Request Size Metrics

Monitoring the Fetch Request Size Avg/Max metrics assists in controlling network load and optimizing fetch performance.

# Consumer I/O Wait Ratio Metric

The Consumer I/O Wait Ratio metric measures the amount of time a consumer thread spends waiting for I/O operations to complete. This is typically time spent waiting for data from disk or over the network. It's exposed in the JMX as `io-wait-ratio-avg` under the `consumer-fetch-manager-metrics` JMX type.

## When the Consumer I/O Wait Ratio Is High

The Consumer I/O Wait Ratio metric refers to the proportion of time the consumer threads spend waiting for I/O operations to complete. A high consumer I/O wait ratio suggests that consumers are spending an excessive amount of time waiting, which can result in slower processing of messages. This could be due to several potential causes, including network latency, disk I/O slowdown, or perhaps an overwhelmed broker unable to respond promptly to fetch requests. A high consumer I/O wait ratio might slow down the rate at which messages are consumed and processed, hampering the overall efficiency of your Kafka consumer.

## When the Consumer I/O Wait Ratio Is Low

A low value implies that the consumer is busy and efficiently utilizing its time.

## Mitigating the Consumer I/O Wait Ratio

When dealing with a high Consumer I/O Wait Ratio, certain fetch parameters can be modified for potential improvement. The `fetch.min.bytes` parameter, for instance, dictates the broker's minimum data threshold that must be met before responding to a fetch request. Increasing this value can make the consumer wait longer before receiving a response, potentially reducing the frequency of I/O operations. However, this waiting time is based on data availability and might not strictly count as I/O wait in the traditional sense.

Similarly, the `fetch.max.wait.ms` parameter, which determines the maximum delay before the broker answers a fetch request when sufficient data to satisfy `fetch.min.bytes` isn't available, can be increased when you have highly fluctuating data production rates. This modification allows fetch requests to wait a little longer for data availability, thereby reducing I/O wait occurrences.

In addition, adjusting the `max.partition.fetch.bytes` parameter, which sets the maximum data returned per partition, can also impact the consumer I/O wait ratio. This parameter essentially governs the amount of data fetched from multiple partitions concurrently. By limiting this value, data fetches from each partition occur more frequently, potentially minimizing I/O wait time.

However, you should remember that these adjustments require careful calibration. While increasing `fetch.min.bytes` and `fetch.max.wait.ms` might diminish the consumer I/O wait ratio, they can also inadvertently increase latency due to prolonged consumer waiting time for data. Likewise, decreasing `max.partition.fetch.bytes` can trigger more fetch requests, thereby raising network and CPU utilization. Therefore, it's essential to strike a balance that satisfies both workload and performance requirements.

### Importance of the Consumer I/O Wait Ratio

Monitoring the Consumer I/O Wait Ratio metric is useful in understanding consumer efficiency and identifying under-utilization issues.

## Records per Request Avg Metric

The Records per Request Avg metric shows the average number of records fetched per request. It's exposed in the JMX as `records-per-request-avg` under the `consumer-fetch-manager-metrics` JMX type.

### When the Records per Request Metric Is High

A high value implies that each request made by a Kafka consumer to a Kafka broker is returning a large number of records.

For consumers, a high average number of records per request could mean that they are efficiently fetching data, especially if large fetches do not lead to a slowdown in processing the records. However, if the consumer cannot process this amount of data efficiently, it might lead to higher memory usage or slower processing time.

On the broker's side, a high number of records per request means that it's serving a substantial volume of data with each request. If this data is cached in RAM (page cache), then the I/O impact is minimal. However, if the data needs to be retrieved from disk because it's not in the cache, then it can lead to increased I/O operations, potentially resulting in longer response time, especially if there's also a high request rate.

Potential causes can be a high rate of message production or a high fetch size configured in the consumer, which allows for more data to be fetched in each request.

### When the Records per Request Metric Is Low

A low value may indicate inefficient fetch operations.

### Mitigating the Records per Request Metric

Adjust the `fetch.max.wait.ms` and `fetch.min.bytes` parameters to influence batching efficiency. For more information on tuning these parameters, check out the section that describes how to deal with a high consumer I/O wait ratio.

### Importance of the Records per Request Metric

Monitoring the Records Per Request Avg metric helps optimize fetch operations and throughput.

## Fetch Latency Avg/Max Metrics

The Fetch Latency Avg/Max metrics reflect the average and maximum time taken to fetch data from the broker.

### When the Fetch Latency Metrics Are High

High average or maximum fetch latency indicates that consumers are experiencing delays when fetching messages. This can be due to network congestion or latency, slow or overloaded Kafka brokers, or potentially from fetching large volumes of data. High fetch latency can lead to lag in the processing of the consumer, as the consumer needs to wait longer to receive the messages before it can process them, affecting the real-time performance and throughput of your Kafka consumer.

### When the Fetch Latency Metrics Are Low

A low value implies swift and efficient fetching of data, suggesting an optimized system.

### Mitigating the Fetch Latency Metrics

If values are high, you should investigate potential network problems, broker performance, and adjust fetch sizes.

### Importance of the Fetch Latency Metrics

Monitoring the Fetch Latency Avg/Max metrics can help you identify bottlenecks and ensure timely data fetches.

## Consumer Request Rate Metric

The Consumer Request Rate metric signifies the number of requests sent by the consumer to the broker per second

### When the Consumer Request Rate Metric Is High

A high Consumer Request Rate metric implies that a consumer is making a large number of requests to fetch data from the broker.

When the consumer request rate is high, it could lead to increased network traffic and potentially overload the broker with incoming requests, causing slower response time and potentially increased latency in message delivery. From the consumer's perspective, a high request rate can also be an indication of increased load or high message consumption rate.

Potential causes for a high consumer request rate could be a high message production rate, leading to more data available for the consumer to fetch. Additionally, the consumer configuration can also influence the request rate. If the `fetch.min.bytes` configuration is set to a small value, the consumer will make more frequent requests.

### When the Consumer Request Rate Metric Is Low

A low value may imply under-utilization of the consumer's capacity or potential network problems.

## Mitigating the Consumer Request Rate Metric

If you're seeing a high consumer request rate in your Kafka cluster, there are a few things you can change to bring it down. First, increase the value of `fetch.min.bytes`. This makes the server wait until there's more data before responding to a fetch request, which can reduce the number of requests your consumer makes.

Next, increase `fetch.max.wait.ms`. This is the maximum amount of time your consumer will wait for a response to a fetch request if there's not enough data to send right away. If you make this time longer, your consumer can wait a bit longer for data to be ready, which can also reduce the number of fetch requests it makes.

Finally, decrease `max.poll.records`. This is the maximum number of records your consumer will try to get in one poll. If you make this number smaller, your consumer will poll more often, but each poll will be smaller, which can help reduce the consumer request rate.

## Importance of the Consumer Request Rate Metric

Monitoring the Consumer Request Rate metric is essential for understanding consumer activity and optimizing performance.

# Bytes Consumed Rate Metric

The Bytes Consumed Rate metric in Kafka indicates how rapidly consumers are extracting data from the Kafka brokers.

## When the Bytes Consumed Rate Metric Is High

A high value for this metric can suggest that consumers are efficiently processing data, but it could also indicate pressure on network and storage resources. It might be due to an unusually high number of consumers in the cluster or some consumers drawing large amounts of data in a short duration. To alleviate this, you can adjust consumer configurations or enhance infrastructure to handle higher data consumption rates.

## When the Bytes Consumed Rate Metric Is Low

On the other hand, a low bytes consumed rate might imply under-utilization of resources, possibly due to slow consumers, network bottlenecks, or inadequate producer data generation. In such cases, it's crucial to identify and address these underlying issues.

## Mitigating the Bytes Consumed Rate Metric

If the consumers are processing data too fast, they can overload the brokers, which can cause problems in the data flow and make the whole system less efficient and reliable.

To manage this, you can adjust the `fetch.max.bytes` parameter, which limits how much data a consumer can fetch at one time. By reducing this, you slow down the rate of processing and make sure your consumers are not overwhelming the brokers.

Sometimes, even with this setting adjusted, the consumers might still be operating too fast. In these situations, you can use a Kafka feature called *throttling*. Throttling works by setting a maximum limit on the amount of data that a consumer or a group of consumers can use. This can prevent any one part of the system from monopolizing the data flow and creating imbalance.

To implement throttling, you can use Kafka's quota API. This lets you set specific limits for each consumer or consumer group by setting the desired data consumption rate.

## Importance of the Bytes Consumed Rate Metric

Monitoring the Bytes Consumed Rate metric helps balance data flow in your Kafka cluster, ensuring optimal utilization of resources and maintaining system performance.

# The Relationship Between Data Skew in Partitions and Consumer Lag

In Kafka clusters, several combinations of consumer lag and data skew can occur among the partitions of the topic the consumers consume. Understanding these scenarios and knowing how to handle them can greatly improve the health of your data streaming pipeline.

If there's no consumer lag and no data skew among the partitions, this is an ideal scenario. Consumers are keeping pace with producers, and data is evenly distributed across all partitions. In such a case, consistent monitoring is essential to ensure the system remains stable and performant.

In contrast, data skew among partitions can sometimes emerge even when there's no consumer lag. Often, the root cause can be attributed to an uneven distribution of partition keys. The *partition key,* a specific data piece in a Kafka message, dictates which partition the message should be directed to. By ensuring messages with identical keys land on the same partition, Kafka not only maintains order but also leverages the similarities between messages for better compression efficiency.

If you observe that the distribution of these keys is uneven, it is beneficial to investigate the producer's key generation and distribution methodology. For situations where the keys aren't paramount, opting for a round-robin partitioner can balance the message distribution.

However, adopting the round-robin method might sacrifice compression efficiency, because it doesn't group similar messages. Grouping messages by their partition key allows for better compression, thanks to their shared content. Therefore, while round-robin can mitigate the risk of data skew compared to partition key usage, it's not always the optimal strategy.

Another limitation of the round-robin approach is its potential impact on the aggregation ratio. When consumers rely on aggregating specific message values, the broad distribution inherent to round-robin can hinder their efforts. For consumers emphasizing such aggregations, especially from a skewed topic, refining the partition key may be more beneficial than defaulting to the round-robin distribution.

When some consumers lag yet no data skew is present among the partitions, it could mean that certain consumers are slower or overwhelmed. Slower consumers can be caused by issues such as garbage collection pauses, slow processing logic, or resource constraints. Consider tuning the number of threads per consumer, adjusting the size of fetched data, or scaling out the consumer application.

A complex scenario arises when some consumers lag and there's data skew among partitions. This indicates that not only are certain consumers slower, but specific partitions may also have more data than others. In such cases, the strategies mentioned previously need to be employed simultaneously.

In cases where all consumers lag, but there's no data skew among partitions, the problem could lie with the consumer application itself or the infrastructure. The consumer applications might be struggling with issues such as garbage collection, resource contention, or slow processing logic. Alternatively, there could be infrastructure problems affecting network performance or disk I/O. Here, scaling out the consumer applications, improving consumer logic, or increasing consumer system resources might be necessary.

The most challenging situation is when all consumers lag and there's data skew among partitions. In such case we need to investigate both the producer key distribution logic to address data skew and consumer application or infrastructure issues to tackle the consumer lag.

Across all scenarios, a comprehensive monitoring setup is key. Using monitoring tools that provide visibility into Kafka cluster metrics, consumer group lag, and partition offsets is essential. Additionally, leveraging logs and application performance monitoring on the producer and consumer applications can help diagnose and rectify these issues more effectively.

To keep an eye on the volume of incoming data across all partitions of a specific topic, you could establish a monitoring strategy for partition skew. This method would display a graph with each line representing the inflow of messages for each partition. The graph's lines consist of individual dots, where each dot signifies the count of events per partition on a minute-by-minute basis.

In this scenario, the P1 percentile corresponds to the partition that receives the smallest quantity of incoming events from all partitions belonging to that particular topic. Meanwhile, the P50 percentile marks the partition receiving the median volume of events. Lastly, the P99 percentile pertains to the partition that gets the most substantial influx of events.

If discrepancies arise between these various percentiles, it is indicative of a data skew concerning the incoming data for this topic.

Here are two examples—one of a topic without partition skew and one with partition skew.

Figure 10-3 shows the number of messages each partition of a specific topic receives per minute. The left Y axis represents the number of messages per minute for a specific partition, and the right Y axis represents the percentile of each partition. You can see that all partitions receive almost the same amount of messages per minute.

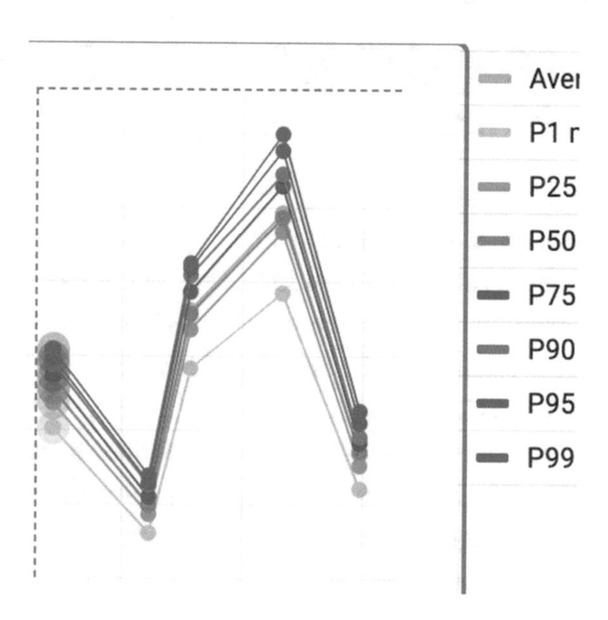


***Figure 10-3.*** *A topic with almost no data skew between its partitions*

Figure 10-4 shows a different story—the partitions from the percentiles of P95 and above receive significantly more messages than the other partitions, which causes their consumers to lag, and potentially even lose data if the lag increases for a longer period of time than the topic's retention.

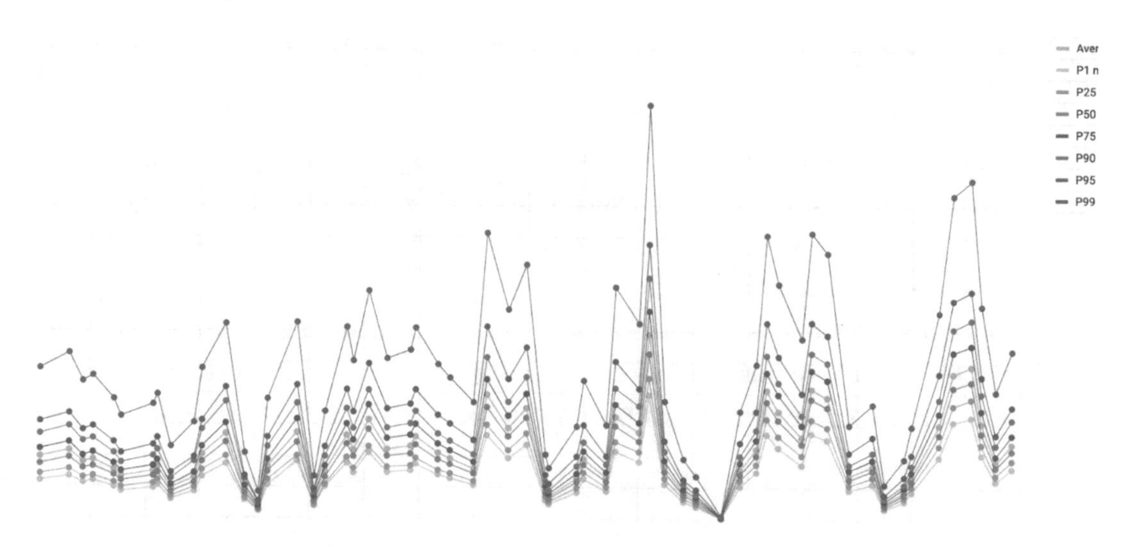


***Figure 10-4.*** *A topic with a data skew. At least two of its partitions (the P95 and P99 partitions) receive more messages than its other partitions*

# An Example of Correlating Between Consumer and Producer Metrics

This section explains how correlating between consumer, producer, and broker metrics can assist you in tracking down the root cause for a rogue broker. I handled a cluster whereby one of its brokers received much more write operations than the other brokers. I determined that by correlating the following metrics—io wait, network processor idle%, avg queue depth, produce latency, fetch consumer latency and log flush rate time.

To begin with, a high `%wa` CPU in the rogue broker (as seen in Figure 10-5) often indicates that the CPU is frequently waiting for I/O operations to complete. In the context of a broker experiencing frequent writes, this suggests that the disk I/O subsystem is finding it challenging to keep pace with the demands of writing data, causing increased CPU wait times.

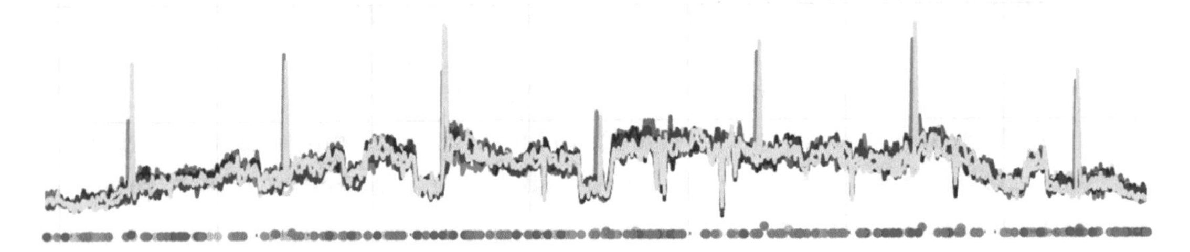

***Figure 10-5.*** *Spikes of high I/O wait time in the rogue broker (represented by the green line)*

Similarly, the network processors can be heavily engaged in managing the influx of incoming write requests. This activity level is reflected in an elevated Network Processor Busy% metric, especially in the rogue broker, as seen in Figure 10-6, which shows the idle% of the network processors.

A drop in the idle% of the network threads means a spike in their busy%. Essentially, every write operation requires network communication, which, when performed at a high frequency, keeps network processors busy, which is reflected in the metric.

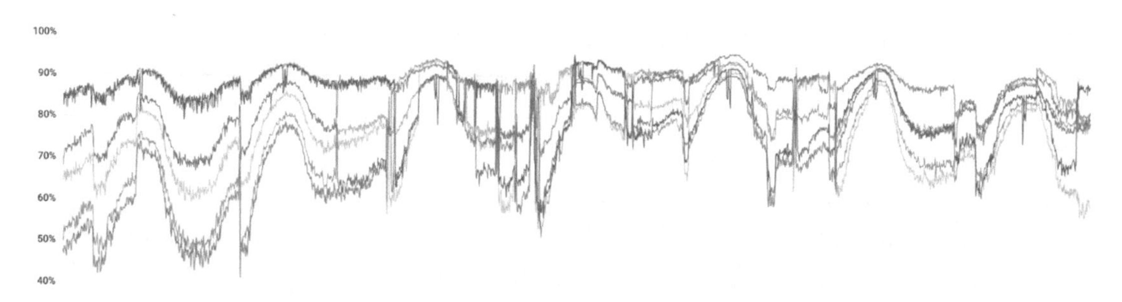


***Figure 10-6.** Drops of network processor idle% time (which is equivalent to spikes in their busy%) in all brokers but especially in the rogue broker (represented by the orange line)*

As the broker grapples with a high volume of write operations, its disks become too busy, so the queue depth (the queue of pending I/O operations) grows, which causes the time that operations spend in the queue also to increase. Figure 10-7 shows the spikes in the number of write operations waiting in the queue in order to perform writes to the disks of the broker.

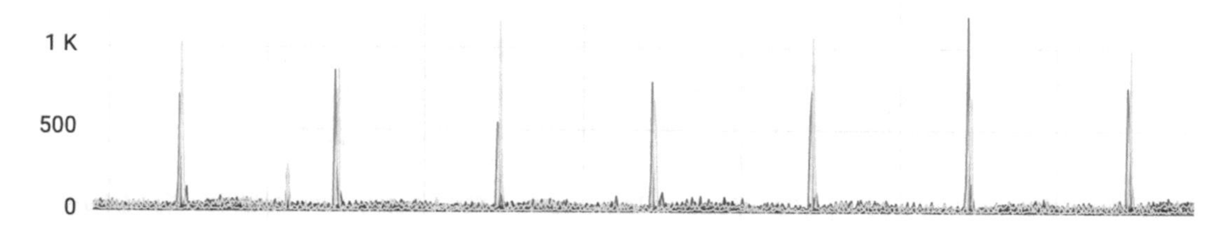


***Figure 10-7.** Spikes in the number of write operations waiting in the queue of the brokers. The rogue broker is represented by the green line*

Furthermore, the Produce Latency 99th Percentile metric captures the longer durations the broker takes to acknowledge write operations under this increased load. Higher latency is a common side-effect of the broker striving to manage high-frequency writes. Figure 10-8 shows the spikes of this metric in the rogue broker.

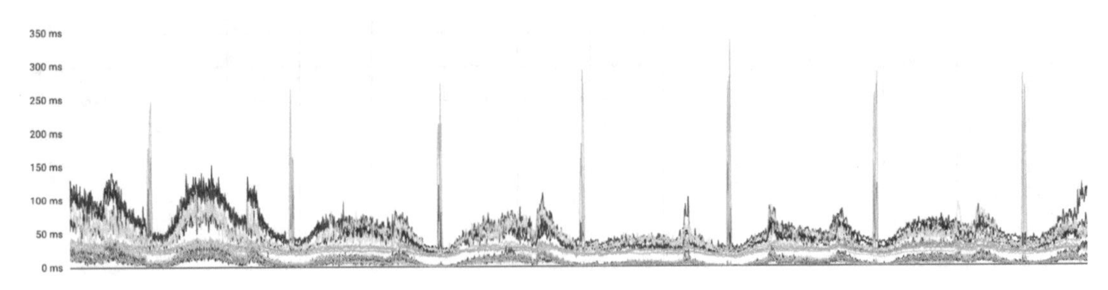


***Figure 10-8.** Spikes of the produce latency 99th percentile in the rogue broker (represented by the pink line)*

In tandem with juggling write operations, the broker also has to serve fetch requests from consumers. As the load increases due to high-frequency writes, serving these fetch requests may get delayed, which in turn increases fetch consumer latency in the rogue broker, as can be shown in Figure 10-9.

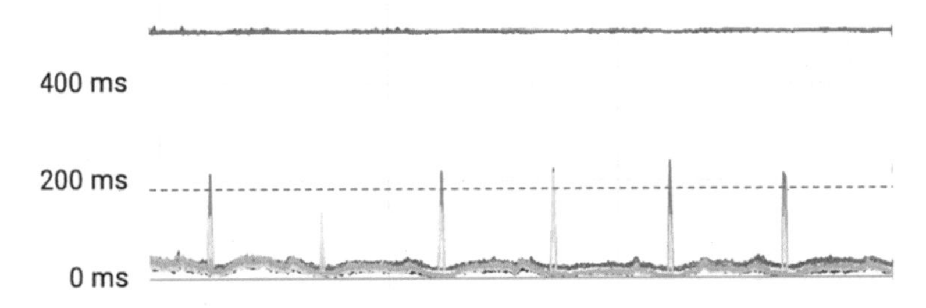


***Figure 10-9.** Spikes in the time it takes the rogue broker (represented by the green line) to serve fetch requests from the consumers*

This Log Flush Rate metric refers to the frequency and amount of time it takes to flush data from Kafka's in-memory log buffer to the disk. This metric increases with frequent writes, as data persistence demands more regular flushes of the log to the disk. Figure 10-10 shows the spikes in the log flush rate in the rogue broker.

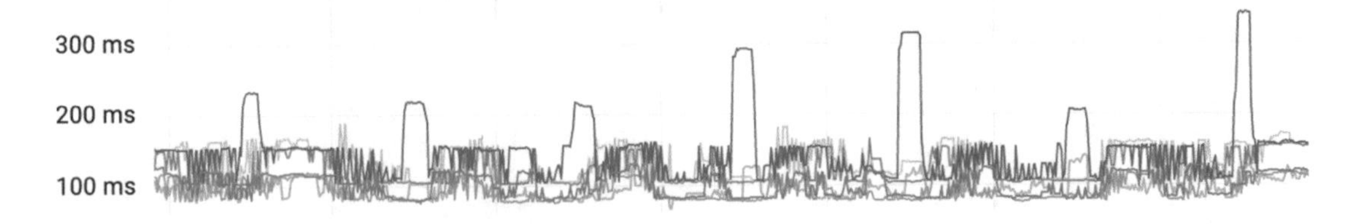


***Figure 10-10.** Spikes in the log flush rate in the rogue broker (represented by the red line)*

To summarize, I used various metrics related to the broker, consumer, and producer to get to the root of the problem with an overloaded broker. By looking at these metrics together, I could see that the broker was dealing with too many write operations. This showed up in metrics like increased CPU wait time, busier network processors, growing queue of operations waiting to be written to the disk, and slower time to acknowledge write operations.

I also saw that the broker was slower to handle fetch requests from consumers and had to flush data more often from its temporary memory storage to the disk. Understanding how these metrics interact helped me figure out why this specific broker was slower. This should remind you that it's important to share the load evenly across all brokers in a Kafka cluster to avoid putting too much strain on one broker.

# Summary

This chapter explained Kafka consumer monitoring. It emphasized the importance of tracking key consumer metrics and drawing insights from their behavior to ensure smooth data flow in the Kafka cluster. The main objective was to enhance the proficiency of consumers in receiving and processing messages, which is pivotal in maintaining the robustness and reliability of the Kafka cluster.

The chapter detailed various consumer metrics, such as Consumer Lag, Fetch Request Rate, Consumer I/O Wait Ratio, and Bytes Consumed Rate, among others. These metrics provide essential insights into aspects like message consumption rates, latency, ensuring the smooth functioning of consumers, and ultimately, the Kafka brokers.

The relationship between data skew in partitions and consumer lag was also explored—understanding different combinations of consumer lag and data skew across the topic's partitions can greatly enhance the health of the data streaming pipeline. An ideal scenario is when there's no consumer lag and data is evenly distributed across all partitions. However, effective management is required when this is not the case.

The chapter also illustrated an example of how correlating between consumer, producer, and broker metrics can help track down issues, using a Kafka cluster as a case study. In this cluster, one broker received significantly more write operations than others. By correlating metrics related to the broker, consumer, and producer, the chapter highlighted how it was possible to identify and resolve the issue with the overloaded broker. The case study underscored the importance of evenly sharing the load across all brokers in a Kafka cluster to avoid overburdening any single broker.

The next chapter delves into the stability of on-premises Kafka data centers. Chapter 11 explores the various hardware components in a Kafka data center, identifying the risks and potential failures that can influence the stability of the system. From disks and RAM DIMMs failures to the challenges posed by Network Interface Cards (NICs), Power Supplies, Motherboards, and Disk Drawers or Racks, we will examine the multifaceted elements that ensure the smooth running of a Kafka cluster. The chapter breaks down the common causes for these hardware failures, the consequences they can have on a Kafka broker, and the strategies that can be implemented to minimize their impact. A special emphasis is on HDD disk failures, given their critical role in Kafka clusters, along with a look at often-overlooked elements like power supplies and motherboards. Additionally, the next chapter discusses how external factors like enabling firewalls and antiviruses can affect performance.

CHAPTER 11

# Stability Issues in On-Premises Kafka Data Centers

Kafka clusters can be separated into two deployment categories—cloud-based and on-premises. This chapter discusses the potential stability issues that may arise from hardware failures in Kafka clusters that are deployed on-premises. Such clusters are heavily dependent on their hardware components and might experience stability issues due to failures in these components. This can impact the cluster stability.

These hardware components specifically include disks, DIMMs, CPUs, network interface cards (NICs), power supplies, cooling systems, motherboards, disk drawers, and cabling and connectors. These components can experience failures for a variety of reasons, including natural wear and tear, manufacturing defects, environmental factors, and improper maintenance.

These failures can significantly degrade the overall performance and stability of the cluster, so we'll investigate the reasons behind these hardware failures, ranging from aging and environmental conditions to manufacturing defects and maintenance issues. Here are some of the effects of these failures:

- Disk failures, either complete or subtle ones like latency and I/O errors, can disrupt Kafka's operation due to its heavy reliance on disk I/O operations.
- DIMMs are central to Kafka's in-memory operations, and their failures can lead to crashes or data corruption.
- CPU failures can reduce the throughput of the Kafka broker, slowing down message delivery. NIC failures can disrupt broker communication, causing delays or data loss.




E. Eldor, *Kafka Troubleshooting in Production*, https://doi.org/10.1007/978-1-4842-9490-1_11

- Power supply failures can lead to unexpected shutdowns, and cooling system failures can cause thermal shutdowns, both disrupting the operation of Kafka brokers.
- Motherboard failures can lead to complete system failure, shutting down the Kafka broker, and failures in disk drawers can cause multiple simultaneous disk failures.
- Lastly, failures in cabling and connectors can cause disruptions in data transmission, network connectivity, and power supply.

Given that these clusters are Linux-based, this chapter explores how to utilize Linux tools to monitor the health of these hardware components, with a particular emphasis on disk monitoring. Such tools can provide insights into disk performance, helping administrators identify potential issues before they escalate.

We'll also discuss the impacts of these hardware failures, including data loss, disruption of replication protocols, and data distribution imbalances, all of which negatively affect Kafka cluster performance.

Understanding hardware failures and their effects on Kafka cluster stability, as well as how to use Linux tools for disk monitoring, will equip you with the skills to prevent, identify, and resolve these issues. The aim of this chapter is to simplify the task of maintaining a stable on-premises Kafka cluster and make it more effective.

# Common Failures in Hardware Components

Hardware components in an on-premises Kafka cluster can experience failures for a variety of reasons, including natural wear and tear, manufacturing defects, environmental factors, and improper maintenance. Here are the key hardware components that are prone to failures:

- *Disks:* Disks are central to Kafka's operation, as they store all of the incoming data. Issues can include mechanical failures, firmware bugs, or problems caused by physical shock or environmental factors. Failures can be complete, preventing access to all data, or partial, causing increased latency or errors during data read/write operations.

- *DIMMs (memory):* Memory is crucial for Kafka's in-memory operations and buffering of data before it's written to disk. Issues with memory can lead to crashes, performance degradation, or data corruption. Memory errors can be transient (a one-time error), intermittent (errors at irregular intervals), or solid (persistent errors).
- *Network Interface Cards (NICs):* NICs are responsible for data transmission between Kafka brokers and other components inside and outside the Kafka cluster. Failures can cause network slowdowns, loss of connectivity, and data corruption.
- *Power supplies:* A failure in the power supply can cause an unexpected shutdown of the hardware, leading to potential data loss or corruption and service unavailability.
- *Motherboards:* A motherboard hosts and interconnects all the hardware components. A failure here can cause the entire system to fail or malfunction.
- *Disk drawers/racks:* Drawers and racks house multiple disks. Power supply issues, improper cooling, and physical damage can lead to multiple disk failures simultaneously.

# The Effect of Hardware Failures on the Stability of On-Prem Kafka Clusters

This section elaborates on the effect of a failure in each of the hardware components that were described in the previous section on the stability of the Kafka cluster.

The impact of these failures can be mitigated by employing replication, failover, and backup strategies, as well as by proactively monitoring the health of the Kafka cluster and its hardware components.

- *Disks:* Disks store all Kafka messages. A disk failure could lead to loss of data if not properly replicated. Disk latency or I/O errors can lead to slow message processing, thereby increasing the end-to-end latency of messages.

- *DIMMs (memory):* Kafka uses memory for buffering data before writing it to disk, and also for caching messages. A failure of a DIMM could lead to increased disk I/O, as more data needs to be read from disk, which could slow down message delivery.
- *Network Interface Cards (NICs):* NIC failures can disrupt the communication between Kafka brokers and between brokers and producers/consumers. This can cause delays in message delivery, replication, and in some cases, can cause data loss if the replication hasn't been completed for some messages.
- *Power supplies:* A failure in the power supply can lead to an unexpected shutdown of a Kafka broker, leading to disruption in service. Messages in the process of being written to disk could be lost, and consumers/producers connected to that broker would be disconnected.
- *Motherboards:* A motherboard failure is catastrophic, as it can lead to a complete system failure, shutting down the Kafka broker and leading to service disruption. If the failure isn't detected and fixed quickly, it could lead to prolonged service disruption.
- *Disk drawers/racks:* Failures here can cause multiple disk failures simultaneously. This can lead to data loss if the disks contain unique data which isn't replicated elsewhere, and can cause a significant increase in disk I/O on the remaining disks as they take over the data handling of the failed disks.

# HDD Disk Failures

One of the most common hardware problems in any on-prem cluster that works heavily with disks is disk failures. Kafka clusters in particular are not only affected by disk failures but are also more prone than other clusters to such failures because they work so heavily with their disks.

There's a higher chance of hardware failures in disks that are used more frequently, especially when the disks are under heavy load or performing a large number of read/write operations, such as Kafka clusters. That's because the mechanical components of the disks are more likely to wear out over time with extended usage, which can eventually lead to hardware failures.

This section discusses reasons for disks to fail and the effects of such failures. Note that I'll refer only to 2.5-inch and 3.5 inch HDD disks, which are connected to Kafka brokers in either SATA or SAS interfaces which spin at 5400-15000 RPM, since these are the only disk types that I've had experience with in Kafka on-prem clusters.

## Common Reasons That Disks Fail

There are several common factors that can cause a disk to fail or to disfunction:

- *Wear and tear:* A disk may wear out due to the constant reads and writes.
- *Power surge:* Data centers may encounter power surges from time to time, due to equipment failures, severe weather, or just electricity being shut down in their region. Some data centers may even lack sufficient backup generators and UPS (uninterruptible power supplies), which prevent them from accessing power until the electricity gets back.
- *Bad sectors:* Bad sectors on a disk are typically created due to wear and tear on the disk surface. They're caused by over-aging of the disks, over-heating, or a filesystem error.

## Potential Impacts of Disk Failure on a Kafka Broker

Consider the potential impact of disk failure on a Kafka broker:

- *Data loss:* If the failed disk contained partitions for one or more topics, data stored on those partitions may be lost and cannot be recovered.
- *Increased latency:* The broker may slow down as it tries to recover the partitions and re-replicate the data to other brokers in the cluster.
- *Reduced throughput:* The broker may become a bottleneck for the cluster as it struggles to keep up with incoming traffic and the increased load from recovery and re-replication.

- *Cluster imbalance:* The failure of a disk on a broker can cause an imbalance in the distribution of partitions across the brokers in the cluster, potentially leading to further performance degradation.

It's important to have a proper backup and disaster recovery plan in place to minimize the impact of a disk failure in a Kafka broker and to ensure high availability of the cluster.

## HDD Disks Lose Their Write Capability and Become Read-Only

Disks can lose their write capability due to a variety of reasons, including physical damage, wear and tear over time, faulty firmware, and software issues. Let's delve into some of these reasons:

- *Physical damage:* Disks can sustain physical damage through excessive heat, water exposure, electrical surges, or any form of physical impact. Overheating may harm electronic components, while a physical collision could cause misalignment or damage to the read/write head or disk platters.
- *Wear and tear:* Hard disk drives (HDD) are particularly vulnerable to wear and tear since they contain moving parts. Over time, these components can deteriorate, hindering the disk's ability to read or write data.
- *Bad sectors:* As disks age, certain areas (sectors) may become faulty and lose the ability to store data. While having a few bad sectors is considered normal, a significant number can signify a failing drive.
- *Full disk:* A disk that has reached its storage capacity will not have space available for writing additional data, effectively rendering it read-only.
- *Incorrect mount options:* The way a disk is mounted can have a direct impact on its write capabilities. If the mount options are set incorrectly, it can prevent users from writing to the disk. This

highlights the importance of properly configuring the mount options, defined in the /etc/fstab file, as part of the Linux filesystem configuration. Ensuring the accuracy of these settings is a crucial step before starting Kafka brokers.

## Monitoring Disk Health

Modern hard drives are equipped with self-monitoring, analysis, and reporting technology (SMART), a vital feature that can provide early warning signs of disk failure. Regular monitoring of SMART data helps you uncover potential issues before they escalate into data loss.

The SMART tool can be utilized to inspect disk devices using the following command, just remember to substitute /dev/sda with the path of the disk you need to inspect:

```
sudo smartctl -a /dev/sda
```

One indication of disk failure from the SMART tool might be:

```
SMART overall-health self-assessment test result: FAILED!
```

For Kafka systems, you can run this command on all devices to detect whether a disk is about to fail:

```
smartctl -a -d megaraid,0 /dev/sdb | grep Health | awk '{print $NF}'
| grep OK
```

Additional verification information from the SMART tool might include:

- Current disk drive temperature
- Disk vendor (useful for replacing faulty disks)
- Disk serial number
- Number of hours powered up
- Total uncorrected errors, indicating the total blocks with uncorrected data errors
- Elements in the grown defect list

In the case of write failures, system logs (e.g., dmesg or /var/log/syslog) may reveal I/O errors.

For those deploying on-premises servers, hardware-specific event monitoring tools like Dell's iDRAC Event Monitor or HP's iLO servers can oversee the health of the system's hardware, alerting administrators if any issues arise. These tools operate similarly, providing invaluable insights into the disk's health.

Kernel messages also serve as an additional approach to identify faulty or bad disks, since they might pinpoint problems that event monitoring tools overlook.

# Remedying Failed Disks

In a Kafka cluster which is deployed on-premises, disk failures can be not only disruptive but also potentially catastrophic, necessitating rapid and effective responses.

When detecting a disk failure in one of your Kafka brokers, the process generally begins with detecting the issue, utilizing tools like the SMART tool (`smartctl`) to swiftly identify if a disk is failing or faulty.

If the disk isn't completely dead yet, you may be able to back up as much data as possible. However, in a production cluster with large disk capacities, backup might be unnecessary and even expensive. When a replication factor of 3 is in place, Kafka's replication feature ensures that data is not lost, making the backup process less critical.

Furthermore, the `smartctl` tool can be used to identify the specifications of the failing disk, information that's vital when choosing a suitable replacement.

Once this assessment is complete, the next step is to replace the faulty hard disk with a new one, ensuring that the new disk meets or exceeds the specifications of the old one.

Once the disk is replaced, format it using a filesystem (such as ext4 or xfs on Linux). Then mount the new disk in the appropriate directory so that it can be used by Kafka. Finally, restart the Kafka broker.

If the failed disk caused an imbalance in the distribution of the partitions across the brokers, it's better to manually reassign the partitions using Kafka's partition reassignment tool. Alternatively, if `auto.leader.rebalance.enable=true`, Kafka will handle it.

After you've replaced the disk and restarted Kafka, monitor the cluster closely for a while to ensure everything is functioning correctly.

You must take specific verification steps after disk replacement, including checking that broker leaders are balanced and assigned to the relevant topic's partitions, confirming that there are no NONE or (`-1`) entries on leaders, validating that all replicas are in sync as shown in the output of the describe tool (in the row with `Isr`), and looking for any errors in the server.log.

While this process can help you recover from a disk failure, preventing disk failures is always preferable. Regular monitoring of disk health (using SMART attributes, for instance) can help reduce the risk of disk failures. Moreover, ensuring that Kafka's data replication is correctly configured can help prevent data loss when disk failures do occur.

# RAM DIMMs Failures

RAM DIMMs can sometimes stop functioning correctly due to various issues, including physical damage, manufacturing defects, and power surges. Errors in RAM can lead to system instability, crashes, and data corruption.

## Potential Causes of DIMMs Failures

Consider these potential causes of DIMMs failures:

- *Physical damage:* Such damage can occur due to mishandling, static discharge, excessive heat or humidity in the server room, or a sudden spike in the power supply (like from a lightning strike or power grid fluctuation). Anything that exceeds the voltage limits of the RAM module's components can cause them to break down or function incorrectly. This is why it's good practice to use a surge protector or an Uninterruptible Power Supply (UPS) with your critical hardware.
- *Age and wear:* Like any component, RAM can degrade over time. Repeated write cycles, in particular, can lead to memory wear.

In most cases, when a DIMM fails, it needs to be replaced. Unlike some components, RAM typically can't be repaired, at least not without specialized equipment and expertise.

## Monitoring DIMMs Failures

Monitoring the health of RAM DIMMs is crucial to maintaining system stability and performance, and recognizing early signs of failure can prevent unexpected crashes and loss of data. Various methods exist to detect and diagnose DIMM issues.

Machine event monitoring, such as Dell's Integrated Dell Remote Access Controller (iDRAC) or HP's Integrated Lights-Out (iLO), offers one such approach. These tools, provided by server manufacturers, continuously monitor the health of hardware components, including RAM. If they detect abnormalities or failures, they can send alerts or log entries. System administrators who keep an eye on these notifications can take preemptive measures before a faulty DIMM leads to serious problems.

Another avenue for monitoring DIMMs comes from the operating system itself. Kernel messages often detect issues with RAM, with error messages related to DIMMs found in system logs, like `dmesg` or `/var/log/syslog` in Linux systems. These messages can include details about specific memory addresses or other technical information, aiding in the diagnosis of the problem.

# Network Interface Cards (NICs) Failures

Network Interface Cards (NICs) are responsible for managing and maintaining the server's connections to other systems. In the context of a Kafka cluster deployed on-premises, NIC failures can lead to serious issues.

## Potential Causes of NIC Failures

Consider these potential causes of NIC failures:

- *Physical damage:* NICs can be damaged through mishandling, static discharge, and overheating.
- *Hardware incompatibility:* Sometimes a NIC might fail due to compatibility issues with the motherboard or other hardware components.
- *Faulty or outdated drivers:* NICs rely on software drivers to function. If these drivers are faulty or outdated, it can lead to failures.
- *Configuration errors:* Incorrect network configurations can cause the NIC to malfunction.

## Implications of NIC Failures on a Kafka Cluster

Issues like these can pop up when you suffer a NIC failure on a Kafka cluster:

- *Loss of connectivity:* When a NIC fails, the broker loses its ability to communicate with other brokers in the Kafka cluster. This can make the broker unavailable, causing client requests to fail and disrupting data processing.
- *Data loss:* If a broker is offline due to a NIC failure and the topic replication factor is low, it could potentially lead to data loss.

## Detecting NIC Failures

There are several Linux tools that can help you diagnose and troubleshoot issues related to NICs. They allow you to understand if the NIC is recognized by the system, if the correct drivers are loaded, and if there are any error messages or other issues affecting the NIC.

- *dmesg:* `dmesg` is a command on UNIX-like operating systems that prints the message buffer of the kernel. It's often used to diagnose issues with hardware, including NICs.
- *lshw:* This is a hardware listing tool that can provide detailed information on the hardware configuration of the system. For NICs, `lshw -class network` will display the configuration, driver, and status of each network interface. This can help identify any NICs that are not working or are not configured correctly.
- *lsmod:* This command shows the status of modules in the Linux kernel. If the driver for a NIC is loaded as a kernel module, `lsmod` can be used to check if that module is loaded. If the module is not listed in the `lsmod` output, that might explain why the NIC is not working.

## Resolving NIC Failures

There are several ways to resolve NIC failures, depending on the kind of failure. If the issue is software-related, you might need to update or reinstall the network driver. You may also need to fix any configuration errors that are causing the issue.

If it's a hardware issue with the NIC itself, consider replacing the NIC if it's a physical card. For built-in NICs, it might be necessary to replace the entire motherboard, or disable the faulty NIC and install a new network card. For redundancy reasons, it's recommended to set two separate network cards for each broker.

# Power Supply Failures

Power supplies are a critical component that can fail, either due to issues with the power supply unit itself or problems with the power source. Such a failure can have significant effects on a Kafka broker, so let's look at the potential causes, implications, and ways to monitor and resolve such failures.

## Potential Causes of Power Supply Failures

Consider these potential causes of power supply failures:

- *Power surges or dips:* A sudden surge in power can damage the power supply unit. Conversely, voltage dips can cause the power supply to fail to provide the necessary power to components.
- *Overheating:* If the power supply's cooling system (usually a built-in fan) fails, the unit can overheat and fail.
- *Component failure:* The power supply unit contains many different components, such as capacitors, which can fail over time.
- *Poor quality or age:* Lower-quality power supply units are more likely to fail, as are older units. Even high-quality power supplies can fail as they age.

## Implications of Power Supply Failures

Issues like these can pop up when you suffer a power supply failure on a Kafka cluster:

- *Unexpected shutdown:* This could potentially lead to data loss or corruption if Kafka is in the middle of a write operation when the power is lost.
- *Service unavailability:* If a power supply fails in a server running a Kafka broker, that broker will go offline. Depending on the replication factor of the Kafka topics, this could lead to service disruption.
- *Hardware damage:* A failing power supply can potentially damage other hardware components in the broker, leading to further issues.

## Resolving Power Supply Failures

Resolving a power supply failure usually involves replacing the failed power supply unit. If you have a redundant power supply, you can replace the failed unit without bringing down the server. Otherwise, you'll need to schedule downtime for the server to replace the power supply.

To avoid disruption due to power supply failures, consider using servers with redundant power supplies, and use a UPS (Uninterruptible Power Supply) to protect against power surges and dips. Also, make sure the server room is well-ventilated to avoid overheating. Regular preventive maintenance can also help detect potential issues before they cause a failure.

# Motherboard Failures

The motherboard is a critical component of any computer system, and its failure can have serious implications.

## Potential Causes of Motherboard Failures

Consider these potential causes of motherboard failures:

- *Power fluctuations:* Sudden power surges or outages can cause damage to the motherboard and other components.

- *Overheating:* If the system cooling is not effective, the motherboard can overheat and potentially fail. This can be caused by a failure of the cooling fan or a buildup of dust.
- *Physical damage:* This could be due to mishandling of the system, for example during transport or maintenance.
- *Component failures:* Failures of other components, especially the power supply, can cause damage to the motherboard.
- *Age:* As with all hardware, motherboards can fail due to age.

## Implications of Motherboard Failures

Issues like these can pop up when you suffer a motherboard failure on a Kafka cluster:

- *System failure:* A motherboard failure can cause the broker to fail, leading to an unexpected shutdown. This can cause potential data loss or corruption and service unavailability. In a Kafka cluster, this would cause one of the brokers to go offline, potentially impacting data availability if the replication factor is not sufficient.
- *Hardware damage:* A failing motherboard can cause damage to other components, leading to further failures.

## Resolving Motherboard Failures

Resolving a motherboard failure usually requires replacing the motherboard, which requires a significant amount of downtime, as it involves disassembling and reassembling the server.

# Disk Drawer and Rack Failures

Disk drawers and racks are physical structures that house multiple disks. They play a crucial role in the organization, cooling, and supplying power to these disks. Failures associated with these components can lead to multiple simultaneous disk failures, which can have serious implications for your Kafka clusters.

## Potential Causes of Disk Drawer and Rack Failures

Consider these potential causes of disk drawer and rack failures:

- *Power supply issues:* If the power supply to the disk drawer or rack fails, all the disks it houses can fail simultaneously. This could be due to a faulty power distribution unit (PDU), power cable, or even a power surge that damages the unit.
- *Cooling issues:* Disk drawers and racks often have built-in cooling mechanisms. If these fail, that can cause the disks to overheat and fail.
- *Physical damage:* This can be from accidents like dropping the rack, water damage, or even simple wear and tear over time.
- *Connectivity issues:* This can be due to faulty cables, connectors, or the failure of the host bus adapter (HBA) that connects the disks to the rest of the system.

## Implications of Disk Drawer and Rack Failures

Issues like these can pop up when you suffer disk drawer and rack failures on a Kafka cluster:

- *Data loss or corruption:* Multiple simultaneous disk failures can lead to data loss or corruption, especially if the replication factor in Kafka is not high enough to ensure data is stored on other brokers.
- *Service unavailability:* Since each Kafka broker typically runs on a separate machine, a full rack failure can lead to multiple brokers going offline, leading to a significant drop in the availability of your Kafka service.

## Resolving Disk Drawer and Rack Failures

Try these methods to resolve disk drawer and rack failures:

- *Replacement:* If a disk drawer or rack fails, it often needs to be replaced. This can require significant downtime, especially if it involves moving multiple disks.
- *Preventive maintenance:* Regular preventive maintenance, including cleaning and physical inspection, can help prevent failures.
- *Good Kafka configuration:* An appropriate replication factor in the Kafka cluster can help mitigate the impact of disk or machine failures.

In summary, while disk drawer and rack failures can have serious implications, proper preventive measures and monitoring can help mitigate the risks associated with such failures.

## Potential Negative Effects of Enabling Firewalls and Antivirus on Kafka Brokers

Antiviruses scans files for malware, often in real-time while files are accessed, created, or modified. Firewalls monitor and control network traffic based on predefined security rules in order to prevent unauthorized access and protect against threats. Enabling antivirus and/or firewall software on the disks of a Kafka broker can potentially introduce latency and affect the consuming and producing rates. Here are some of the effects that they can have on a Kafka cluster:

- *Disk I/O latency:* If your AV is configured to scan the directories where Kafka is storing its log files, it could potentially cause a significant increase in disk I/O operations, as every write to a Kafka topic log file would also involve a read operation by the antivirus software. This added I/O overhead could result in higher disk latency, slowing down the rate at which Kafka can write to or read from its log files. This could in turn affect the latency for Kafka producers and consumers. If possible, it's recommended to configure the antivirus software to scan only at specific times or to exclude the Kafka broker data directories from scanning.

- *System resources:* If the AV is configured to run frequent scans, it can consume a significant amount of system resources and impact the performance of the Kafka broker.
- *Network traffic:* Kafka is a distributed system that relies heavily on network communication. Misconfigured or overly restrictive firewall rules can impede this communication, affecting the performance of the Kafka cluster.
- *Inter-node communication:* Kafka brokers communicate with each other (inter-broker communication), especially in replication scenarios. Firewalls need to be configured to allow this inter-broker communication.
- *Impact on ZooKeeper:* Kafka uses ZooKeeper to maintain and coordinate brokers. Firewall rules should allow traffic between Kafka and ZooKeeper processes.

It's important to carefully consider the impact of firewalls and antivirus software on the performance of a Kafka broker before enabling them. Moreover, sometimes Kafka administrators just don't notice that this software is installed on the Kafka broker, so it's a good practice to develop a verification script that will run once in a while on the brokers and verify whether firewalls and antivirus programs are installed. If they are installed, the program should also ensure that they aren't enabled.

# ZooKeeper Best Practices in On-Premises Kafka Data Centers

The efficiency and stability of an on-premises Kafka data center are inextricably linked to the underlying infrastructure and configuration practices. Among the critical components of this ecosystem is Apache ZooKeeper, a distributed coordination service that plays a pivotal role in managing the Kafka cluster. As it houses the metadata and provides synchronization across the cluster, its reliability is paramount for the proper functioning and stability of the entire Kafka system.

Ensuring that ZooKeeper is optimized, both in terms of hardware configuration and monitoring practices, can significantly contribute to the overall robustness of a Kafka deployment. This section delves into key best practices that cater to the specific needs of ZooKeeper in a Kafka environment, including considerations for disk performance, the importance of dedicated machines, and strategies for continuous monitoring and assessment.

## Disk Performance

Ensuring optimal disk performance is vital for maintaining a healthy ZooKeeper cluster. Solid State Drives (SSDs) are strongly advised for ZooKeeper, as they offer the low-latency disk writes required for optimal functioning. Since each request to ZooKeeper must be committed to disk on every server in the quorum before the result becomes available for reading, having efficient and fast storage is a non-negotiable requirement. Monitoring the I/O performance, disk latency, and write speeds can prevent bottlenecks that might otherwise hamper the overall system performance.

## Dedicated Machines

Another significant recommendation for ZooKeeper deployment is to host the servers on dedicated machines, separate from the Kafka broker cluster. This isolation ensures that ZooKeeper can function at its best without competing for resources with Kafka brokers. Such a setup allows for more precise tuning, monitoring, and maintenance of the ZooKeeper instances, which are crucial for stability.

## Monitoring ZooKeeper

To make sure the Kafka cluster remains stable, you must pay careful attention to several aspects of ZooKeeper. Keeping an eye on the up or down status of the nodes in the ZooKeeper quorum is vital, as is monitoring the response time for client requests to detect performance issues early on. It's also essential to watch the data stored by ZooKeeper, ensuring it stays within healthy limits. Regular checks on the number of client connections to the ZooKeeper servers help you understand the load and potential stress on the system.

## Developing a Dedicated Smoke Test for Kafka and ZooKeeper Stability

In a complex and dynamic environment such as Kafka, staying ahead of potential issues and inconsistencies is essential to maintaining the desired level of performance and stability. One robust way to accomplish this is by developing and implementing dedicated *smoke tests*. Smoke tests are quick preliminary tests that cover essential functions of a system. In the context of Kafka and ZooKeeper, they can be incredibly valuable in catching and addressing problems in a timely manner, before they escalate into more significant challenges.

The idea behind smoke testing in this environment is to create a script that can periodically run a series of checks and validations on the hardware and the Kafka cluster in production. Here's a detailed look at the various validations that you can include:

- *DIMM's issues:* Check for any RAM issues from kernel messages, as these can impact overall system stability.
- *Disk issues:* Analyze kernel messages or use tools like `smartctl` to inspect disks for any problems that might hinder performance.
- *Network/NIC problems:* Utilize kernel messages or dedicated tools like `ethtool` to ensure that the network interfaces are operating correctly.
- *Available memory across Kafka machines:* Ensure that there is enough memory available for smooth Kafka operation.
- *Number of open files:* Check that the number of open files has not reached a critical threshold, which could limit Kafka's ability to function.
- *Kafka disk usage:* Monitor the disk space dedicated to storing Kafka topics across machines to avoid running out of space.
- */root and /var filesystem usage:* Confirm that these filesystems have not reached critical usage levels, which can lead to storage-related issues.
- *Time synchronization:* Ensure all machines are in sync with the NTP server, as inconsistent time-keeping can lead to various issues.

- *Disk balance across Kafka machines:* Check that disk space usage is more or less balanced, to prevent certain machines from being overloaded.
- *Out-of-memory problems on Kafka service:* Monitor the `kafka.err` log for memory-related errors, which can impact Kafka's performance.
- *Service availability:* Confirm that all Kafka broker services and ZooKeeper services are up and running.
- *Max connections on ZooKeeper servers:* Ensure that connections have not reached critical thresholds, especially when ZooKeeper serves other services.
- *Disk mounting and read/write status:* Check that all disks are mounted correctly and that none are in a read-only state, which could limit functionality.
- *Broker IDs in ZooKeeper:* Ensure that broker IDs are appropriately registered in ZooKeeper, which is essential for the correct operation of the cluster.

# Summary

This chapter delved into various factors that can influence the stability of on-premises Kafka data centers. It began by outlining the potential failures in key hardware components. These included disks, RAM DIMMs, NICs, power supplies, motherboards, and disk drawers or racks, which can all experience failures due to natural wear and tear, manufacturing defects, environmental factors, and improper maintenance.

Next, the chapter delved into the impact of these hardware failures on the stability of a Kafka cluster. It highlighted that the consequences of these failures could be lessened by employing strategies such as replication, failover, and backups, as well as proactively monitoring the health of the Kafka cluster and its hardware components.

A particular focus was given to HDD disk failures. We discussed the reasons for these failures and their effects on Kafka clusters. The discussion broke down common reasons for disk failures, their potential impacts on a Kafka broker, the process of monitoring disk health, and ways to resolve disk failure issues.

We also looked at RAM DIMMs failures, learning potential causes such as physical damage or natural wear and tear. We learned about the effects of these failures on system stability and data integrity.

Then, we shifted focus to Network Interface Cards (NICs), which are vital for managing server connections. That section explored how failures in NICs, caused by factors like physical damage, hardware incompatibility, and faulty drivers, can lead to significant problems in on-premises Kafka clusters.

Power supplies, a critical but often overlooked component, were discussed next. That section explored the potential causes for power supply failures, their effects on a Kafka broker, and methods to monitor and resolve such issues.

The next section covered motherboards. As the backbone of a computer system, motherboard failures can have severe implications. That section covered potential causes, impacts, and solutions for these failures.

We then moved on to discuss disk drawers and racks, which house multiple disks. The section illustrated how a failure of any of these components could result in multiple simultaneous disk failures, and we looked into the causes, implications, and ways to mitigate such failures.

We also discussed the potential negative effects of enabling firewalls and antivirus programs on Kafka brokers. We touched upon potential latency issues and the impact on consuming and producing rates that may occur when enabling antivirus and firewall software on the disks of a Kafka broker.

We also explored ZooKeeper's best practices in on-premises Kafka data centers and emphasized optimal disk performance, low-latency writes using SSDs, and the significance of dedicated machines for ZooKeeper. Comprehensive monitoring strategies were outlined, highlighting ZooKeeper's role in Kafka cluster stability.

We concluded with an in-depth look at the development of dedicated smoke tests for maintaining Kafka and ZooKeeper stability. Detailed insights were provided on various tests for hardware and Kafka cluster validations, including RAM issues, disk problems, network functionality, memory availability, disk usage, and more. The section underscored the preventive role of smoke tests in identifying and addressing potential problems early.

The next chapter focuses on an aspect that can significantly impact the efficiency and cost-effectiveness of a Kafka cluster: optimizing hardware resources.

While the stability and robustness of the Kafka cluster are of paramount importance, an equally critical consideration is ensuring that the system is not over-provisioned. Over-provisioning can lead to unnecessary costs and underutilized resources, resulting in an inefficient system.

Chapter 12 dissects the various metrics and considerations to accurately assess the cluster's usage, such as RAM, CPU, disk storage, and disk IOPS. It explores the nuanced differences in scaling on-prem versus cloud-based clusters and investigates the pros and cons of different scaling alternatives from various perspectives, including technical, managerial, and financial. Through six illustrative examples, you will get a detailed guide on how to effectively implement scale-in and scale-down strategies to maximize cost savings without compromising the stability of your Kafka cluster.

# CHAPTER 12

# Cost Reduction of Kafka Clusters

This chapter zeroes in on the pivotal aspect of reducing hardware costs in Apache Kafka clusters. By carefully examining the choices between deploying Kafka on the cloud and on-premises, we'll delve into strategies that can lead to significant cost reduction.

The chapter begins by exploring the vital aspects that influence the scaling of Kafka clusters, with particular attention to Kafka on-premises. Here, we'll outline various options and their direct impact on hardware costs, drawing comparisons between cloud and on-premises solutions. The technical, managerial, and financial facets are analyzed, all within the context of achieving cost savings.

As we progress, the chapter takes a deep dive into the hardware considerations that play a central role in controlling costs. This includes detailed examinations of RAM, CPU cores, disk storage, and IOPS, with an emphasis on identifying the most economical configurations and setups. Practical examples are provided to clarify the optimal hardware selection for both cloud and on-premises deployments, with clear insights into the tradeoffs involved.

The concluding section presents a series of real-world examples that vividly illustrate proven strategies for cost reduction in over-provisioned Kafka clusters. These include specific findings, options for reducing costs, and recommendations tailored to different scenarios and cluster specifications. Special attention is given to cloud deployments, but the principles can be applied more broadly.

Through these examples and the detailed exploration throughout the chapter, you'll gain a concrete understanding of how to minimize the hardware costs of your Kafka cluster. The insights offered are not merely theoretical; they are drawn from real-world applications and are designed to empower you to make informed, cost-effective decisions.




E. Eldor, *Kafka Troubleshooting in Production*, https://doi.org/10.1007/978-1-4842-9490-1_12

# Determining If You Can Reduce Costs in Your Kafka Clusters

When it comes to scaling a Kafka cluster, the deployment environment makes a crucial difference. Specifically, scaling is often much more straightforward when the cluster is on the cloud as compared to on-premises. This difference stems from both technical and managerial reasons that make scaling an on-premises Kafka cluster more complex.

In discussing the scaling of an on-premises cluster, we need to take into consideration three key aspects: technical, managerial, and financial. Scaling the cluster in this context refers to making adjustments that may include adding or removing brokers or changing the type of the brokers. These changes can encompass variations in RAM, CPU cores, disk storage, or even disk type.

To illuminate the distinction between deploying on the cloud and on-premises, the following sections delve into the various reasons for scaling a cluster and explore how this process can be accomplished in both environments.

## Lack of RAM

Managing memory in a Kafka cluster is essential. In the cloud, you can easily add more RAM by choosing bigger instances. But for clusters deployed on-premises, it's a bit more complex. You can either add or swap out memory sticks, or just bring in more machines. Each choice has its own set of pros and cons, so it's crucial to think it through.

### Cloud-Based Cluster

In order to add more RAM to a Kafka cluster that is deployed on the cloud, we just need to spin off a new cluster with new instances that have more RAM.

### On-Prem Cluster

When aiming to add more RAM to a Kafka cluster deployed on-premises, we have several possible approaches. If there are available memory slots on the motherboard, additional DIMMs can be added to each broker. Alternatively, if the size of the current DIMMs deployed on the motherboard is smaller than the maximal size, these can be replaced with larger ones.

If all memory slots are occupied with DIMMs at their maximal size, adding more machines becomes the viable option. It's important to note that the term DIMM used here refers to a memory module or memory stick. This context is specifically referring to DDR4 DIMMs, which range in size from 8GB to 64GB.

## Discussion

There are several courses of action when dealing with a Kafka cluster that's deployed on-prem and lacks RAM, and choosing which way to go depends on the use case. If there are available slots for DIMMs, then the easiest way is to add DIMMs. If that solves the issue, then we're done.

However, what if it doesn't solve our problem? In such case, we have two options, as discussed next.

### Replace the DIMMs with Larger DIMMs That Have More RAM

This strategy can be employed if the current size of the DIMMs isn't at its maximum. Replacing all the DIMMs in all brokers with larger-sized DIMMs has specific advantages and challenges.

From a technical standpoint, this approach is correct, as it merely involves adding RAM without introducing additional brokers. This means there's no need to reassign the topics since the number of brokers remains the same.

However, the financial implications of this solution must be considered. For instance, purchasing a single server with 24 DIMMs of size 32GB may cost between two to five times more than purchasing just the DIMMs themselves, depending on the server type.

The managerial aspect also presents challenges, making it a tough decision to undertake.

This process involves removing all existing DIMMs from the brokers, replacing them with the larger ones, and storing the old DIMMs elsewhere. Data center owners often resist purchasing hardware that will not be utilized (such as the old DIMMs), making this solution less attractive from an operational perspective.

If the current size of the DIMMs isn't the maximal one, we can replace all the DIMMs in all brokers with DIMMs of a bigger size.

### Scale Out by Adding More Brokers with the Same Amount of RAM

This is an alternative solution to increasing the cluster's capacity. From a managerial standpoint, this method has the advantage of not requiring the disposal of old DIMMs, making it an easier decision to implement. Moreover, adding more machines to the cluster is often a simpler task compared to replacing all the DIMMs in the current brokers.

However, this approach does have its drawbacks. If the DIMMs in the current brokers can be replaced with larger ones, adding more brokers with the same amount of RAM might end up being more costly. Essentially, the financial implications could outweigh the convenience, especially if the current DIMMs have not reached their maximum potential size. Therefore, careful consideration of the technical requirements and the budget constraints should guide the decision-making process.

The decision whether to replace the DIMMs or add more brokers needs to take into account all these aspects—technical, financial, maintenance and managerial. In order to make the right call, you'll need to put into your calculation the importance of each aspect, and according to that decide on the correct path.

## Lack of CPU Cores

### Cloud-Based Cluster

When addressing the lack of CPU cores in a Kafka cluster deployed on the cloud, there are two main strategies that we can pursue: scale up or scale out.

- For the scale-up approach, we can create a new cluster with the same number of instances but select instance types that come with more cores. This essentially enhances the existing configuration with additional processing power without increasing the number of instances.

- The scale-out method involves adding more instances (with the same instance type of the the existing instances) to the cluster. This expands the cluster's size without changing the individual processing capabilities of each instance, providing a broader rather than deeper enhancement.

## On-Prem Cluster

When an on-premises Kafka cluster requires more CPU cores, two primary strategies can be considered: scaling up or scaling out.

- The scale-up approach involves creating a new cluster by replacing the existing machines with new ones that have more CPU cores. This strategy has the advantage of ensuring that the cluster will have the necessary cores, but it comes with financial and managerial challenges. For example, the old machines, now unused, represent an additional cost, and explaining to the data center owner why a cluster was initially provisioned with insufficient CPU cores can be challenging.
- The scale-out approach, on the other hand, entails adding more brokers of the same type to the existing cluster. This method avoids the need to remove current brokers, making it more palatable from both a financial and managerial perspective. However, it may lead to other complexities, such as ending up with a cluster that meets CPU requirements but has excessive RAM, disk storage, or disk IOPS.

## Discussion

When facing the decision whether to scale out or scale up a Kafka cluster to get more CPU cores, various considerations come into play, and the optimal approach might differ between on-premises and cloud-based deployments.

In the context of an on-premises cluster, scaling out often seems more financially attractive, as it can extend existing resources without necessitating the purchase of entirely new or costlier hardware. Conversely, in a cloud-based environment, financial considerations may be less significant, as replacing machines doesn't result in the retention of old, unused hardware.

From a managerial perspective, scaling out an on-prem cluster may also be more appealing. It avoids the disposal or replacement of current equipment, aligning more closely with the existing infrastructure and investment, and making the decision-making process smoother. This managerial concern is typically not relevant for a cloud-based cluster, where hardware replacement is a transparent operation.

Technically, scaling out is a more straightforward solution for both on-prem and cloud-based clusters. It allows for the necessary expansion without the need to migrate topics from the old cluster. Reassignment of partitions to new brokers simplifies the process, making scaling out a commonly preferred method across both deployment scenarios.

## Lack of Disk Storage

### Cloud-Based Cluster

When adding more disk storage to a Kafka cluster deployed on AWS, it's crucial to consider the underlying storage technology, namely Elastic Block Store (EBS) and NVMe (Non-Volatile Memory Express) devices. Each of these options has unique characteristics and considerations.

With EBS, you have the flexibility to attach volumes to existing instances, allowing you to increase disk space without altering the existing infrastructure. You can also scale out the cluster by adding more brokers with the same amount of EBS disk space, distributing the data and load across a more extensive set of nodes. If needed, you can scale up the cluster by replacing current brokers with instances that have larger EBS volumes. Since EBS offers various types and sizes, you can choose the best fit for performance and cost, bearing in mind that EBS is network-attached storage, which might have implications on latency and IOPS requirements for your Kafka workload.

On the other hand, NVMe devices, known for low-latency and high-throughput storage, offer different opportunities for scaling. If your instances support NVMe, you can attach additional NVMe disks to each broker. This option may be especially beneficial for write-intensive workloads, giving you the option to scale out by adding more brokers equipped with NVMe disks. Alternatively, you can scale up by selecting instances with larger NVMe storage, allowing for a seamless increase in disk space and potential performance benefits. Keep in mind that NVMe is typically local to the instance, so data durability and replication strategies must be meticulously planned.

### On-Prem Cluster

In an on-premises environment, adding more disk storage to a Kafka cluster requires careful consideration of the existing hardware configuration, along with identifying the most suitable method for expansion. The term *storage drawer,* which refers to the component within a server chassis that houses disk drives, is pivotal in this context.

One strategy to increase storage involves replacing the current disk drives in the storage drawers with those having more storage capacity. This approach enhances the existing infrastructure without necessitating additional hardware, capitalizing on the opportunity to upgrade without substantial changes.

If there are unused slots in the storage drawers, then adding more disk drives can be a viable option. This approach makes the most of existing capacity in the infrastructure, promoting a cost-effective increase in storage without transforming the overall hardware configuration.

Alternatively, the cluster can be expanded by adding more brokers, each equipped with the same number and type of disks.

## Discussion

When confronting the challenge of adding more disk storage to a Kafka cluster, various factors must be examined, and the most suitable approach may differ between cloud-based and on-premises deployments.

In the cloud-based environment, particularly with AWS, the availability of different underlying storage technologies, such as Elastic Block Store (EBS) and NVMe, provides flexibility in scaling. With EBS, you can extend disk space effortlessly, offering options to scale out by adding more brokers or scale up by choosing larger storage volumes. However, considerations regarding network latency and IOPS requirements must be factored in. On the other hand, NVMe devices offer low-latency and high-throughput storage, but data durability and replication strategies must be carefully planned, as NVMe is typically local to the instance.

For an on-premises cluster, the decision-making process requires a thorough understanding of existing hardware configurations. Options might include replacing existing disk drives with larger ones, adding more drives if slots are available, or expanding the cluster with more brokers with the same type of disks. The choices often hinge on financial efficiency and alignment with the existing infrastructure, along with performance requirements. The approach must be consistent with the Kafka workload, ensuring that considerations such as latency, throughput, and data replication are adequately addressed.

From a managerial perspective, decisions regarding scaling disk storage in an on-prem cluster often lean toward maximizing existing resources without unnecessary expenditures on new hardware. This aligns with typical data center ownership concerns, favoring solutions that work with the existing investments. By contrast, the flexibility and variety of options in a cloud-based environment might allow for a more straightforward scaling, but with careful attention to specific storage characteristics and their impact on performance and reliability.

## Lack of Disk IOPS

If the disks have reached their maximum capacity for disk IOPS, there are two options to add more IOPS. The first option is adding more disks of the same type or replace the existing disks with ones that have a higher IOPS. You can find more info on that in the "Lack of Disk Storage" section.

However, if the lack of disk IOPS is due to read operations on the disks, the problem may not be lack of disk IOPS, but rather lack of RAM or lagging consumers. This is because the reads are initially performed from the page cache, and only if the messages are not found there, the operating system will read the data from the disks. In this case, instead of adding more disk IOPS to the cluster, it may be beneficial to check whether consumers lag, and/or add more RAM. You can find more information on this in the "Lack of RAM" section.

# Cost Optimization Strategies for Kafka Clusters in the Cloud

To reduce the hardware costs associated with a Kafka cluster, two main strategies can be considered: scaling in the cluster by using fewer brokers, or scaling down the cluster by using smaller brokers. Each approach has unique benefits and challenges. The following sections explore examples of over-provisioned Kafka clusters, starting with an in-depth description of the cluster's specifications and hardware resource utilization.

Following that, we'll assess the options for scaling in and scaling down the cluster where relevant. We'll also illustrate the potential cost savings for each strategy, measured as a percentage of current costs, and offer guidance on the best scaling method to minimize costs without compromising the cluster's stability.

# Additional Considerations

Before we delve into the specific examples of over-provisioned Kafka clusters, this section goes over several considerations that may influence your decision-making process. From the deployment environment (cloud vs. on-prem) to the technical details such as hyper-threading, CPU types, and normalized load averages, these aspects provide the context for the subsequent analysis.

# Cloud vs. On-Prem

The following examples refer only to Kafka clusters that are deployed in the cloud, and the recommendation of how to reduce the costs of these clusters refers only to how to perform this in the cloud and not on-prem.

## Hyper-Threading

The number of CPU cores in each example is calculated under the assumption that hyper-threading is enabled. When hyper-threading is enabled, the Linux kernel can create two logical processors (threads) for each physical core, allowing two threads to execute simultaneously on a single core.

For example, consider a server that has two sockets. Each socket has 12 cores, and hyper-threading is enabled in the Linux kernel. This server has 24 physical cores but the OS sees 48 cores due to the hyper-threading. In this case, we'll refer to this server as having 48 cores and not 24, since that's the number of cores that the OS sees.

## CPU Type

There are different CPU types available in AWS, such as i386, AMD, and ARM. Each of these types has unique characteristics that may influence the performance and cost of the Kafka cluster. However, the influence that each CPU type has on the Kafka cluster isn't discussed in this chapter.

## Normalized Load Average (NLA)

One of the metrics that we'll use in order to indicate the load on the clusters is the Normalized Load Average (NLA), so let's look at how the NLA is computed.

While the *load average* is a measure of the number of processes that are currently running or waiting to run in the CPU queue (for time periods of the last 1, 5, and 15 minutes), the *normalized load average* is a scaled value that represents the load average relative to the number of CPU cores on the system. The value of the NLA is defined as: (Load Average/Normalization Factor) x 100.

For example, consider a machine with eight cores. If the load average for the past five minutes is four, that means that on average there were four processes either in a runnable or waiting state. In that case, the NLA is (4/8) X 100 = 0.5.

## Scale In

For the sake of clarity, in all the examples that follow, when we explore options to scale in a Kafka cluster, the sections specifically look at using instance types that belong to the same family as the original brokers. While there is indeed the option to scale in a cluster by using instance types from a different family, this chapter doesn't consider that approach, in order to maintain simplicity in this examination.

### Example 1

Table 12-1 shows a Kafka cluster with four brokers.

***Table 12-1.*** *A Kafka Cluster with Four Brokers*

| # CPU Cores Per Broker | RAM per Broker | Disk Storage Per Broker (NVMe) | %us | %sy | %wa |
|---|---|---|---|---|---|
| 48 | 384GB | 30TB (4 disks) | 5% | 20% | 0% |

| Total CPU% Utilization | Normalized Load Average | Used Disk Space in /var/lib/kafka | Network Processor Idle |
|---|---|---|---|
| 25% | 0.25 | 15% | 99% |

## Findings

*CPU utilization:* The cluster is over-provisioned in terms of CPU (only 25% CPU utilization).

*Disk storage usage:* The cluster is over-provisioned in terms of disk storage (only 15% disk storage usage).

*Disk IOPS utilization:* There are no reads from the disks (because the `%wa` is low).

*Load on the Cluster:* The NLA is far from reaching 1.0, which means that the load on the cluster is low.

# Options for Reducing Costs

There are two ways to reduce the hardware costs of the cluster—scale down or scale in.

Let's look at the implications of each scaling option and then compare between the two. This chapter recommends the option that is best in terms of cost reduction and maintenance.

## Scale Down

Assuming that you want to remain with the same instance family, you can replace the current brokers with brokers that have half the cores, disk storage, and RAM. The reason is that in AWS, the next instance type that's smaller than the current instances has 24 cores, 15TB storage, and 192GB RAM. Let's check each hardware aspect and see how much you can scale it down:

- *CPU:* Use 24 cores instead of 48. Since even with half the number of cores, the brokers will have CPU utilization of only 50%, 24 cores seems to be enough, given that you have a rough estimation that CPU utilization is linear to the number of CPU cores. This is a valid estimation since the CPU utilization is mostly contributed to `us%` and `sy%` and not to disk wait time (`wa%`).

- *Storage:* In AWS, instances with 24 cores arrive with half of the disk storage (15TB) compared to instances with 48 cores. With this amount of storage, the storage usage will be only 30 percent, which is still low.

- *RAM:* In AWS, instances with 24 cores arrive with half of the RAM (192GB RAM) compared to instances with 48 cores. You can tell whether the RAM is sufficient only by reducing the amount of RAM and then checking if there are more reads from the disks. Since there are currently no reads from the disks, you can use only 192GB RAM and see if it's enough by checking the rate of read IOPS from the disks.

## Scale In

Since the cluster has four brokers, the option of scaling in the cluster depends on the replication factor (RF) of the topics in the cluster.

If the RF is 3, it's not recommended to remove even a single broker. The reason is that if a single broker fails in a cluster, only two brokers will be left in the cluster and the requirement of three replicas per topic won't be satisfied.

However if the RF is 2, you can remove a single broker and leave the cluster with three brokers. Even if one broker fails, the cluster will still have two brokers, which means that each topic will still have two replicas and the replication requirement will be met.

# Recommendation

Scaling in the cluster would reduce the hardware costs by 50 percent since all the brokers will be replaced with brokers at half the price (and half the hardware resources as well). This will require migrating all the topics to the new brokers without having to change the number of partitions.

On the other hand, scaling down the cluster would reduce the costs by 25 percent, but that's recommended only in case the replication factor of the topics is 2 and not 3. This will require a reassignment of the partitions and a change to the number of partitions of all the topics in which their number of partitions doesn't divide equally by 3.

## Example 2

Table 12-2 shows a Kafka cluster with six brokers.

***Table 12-2.*** *A Kafka Cluster with Six Brokers*

| # CPU Cores Per Broker | RAM Per Broker | Disk Storage Per Broker (NVMe) | %us | %sy | %wa |
|---|---|---|---|---|---|
| 8 | 64GB | 5TB (2 disks) | 30% | 12% | 0% with peaks of 4% |

| Total CPU% Utilization | Normalized Load Average | Used Disk Space in /var/lib/kafka | Network Processor Idle |
|---|---|---|---|
| 45% | 0.6 | 4% | 60% |

## Findings

*CPU utilization:* The cluster is over-provisioned in terms of CPU (only 45% CPU utilization).

*Disk storage usage:* The cluster is over-provisioned in terms of disk storage (only 4% disk storage usage).

*Disk IOPS utilization:* Most of the time, the disks aren't utilized, but during the day there are several spikes of reads from the disks. This causes the CPU `wa%` to reach a value of 4%, which is quite high.

*Load on the cluster:* The NLA is 0.6, which isn't low but also not that high.

*Network processor idle percentage:* This value is really low, only 60%.

## Options for Reducing Costs

A low value for Network Processor Idle indicates some bottleneck in the cluster or that there are just not enough network threads. In a healthy cluster, this value should be 99 percent, and it's surprisingly low given the fact that both CPU usage and load average aren't that high. The spikes in the `wa%` also indicate there's some issue in the cluster that causes the disks to be highly utilized in terms of disk IOPS.

## Recommendation

At first sight, this cluster seems over-provisioned in terms of CPU cores, since its CPU utilization is at ±45%. But the low network processor idle option shows that it won't be a smart move to scale down the cluster because it suffers from some issue that might already cause latency for its clients, and reducing cores might make these symptoms worse.

So in this case you should focus on investigating the cause of the problematic symptom instead of reducing the cost of the cluster.

### Example 3

Table 12-3 shows a Kafka cluster with 12 brokers.

*Table 12-3. A Kafka Cluster with 12 Brokers*

| # CPU Cores Per Broker | RAM Per Broker | Disk Storage Per Broker (NVMe) | %us | %sy | %wa |
|---|---|---|---|---|---|
| 12 | 96GB | 7.5TB | 27% | 12% | 2% |

| Total CPU% Utilization | Normalized Load Average | Used Disk Space in /var/lib/kafka | Network Processor Idle |
|---|---|---|---|
| 40% | 0.5 | 50% | 99% |

## Findings

*CPU utilization:* The cluster is over-provisioned in terms of CPU (only 40% CPU utilization).

*Disk storage usage:* The cluster is over-provisioned in terms of disk storage (only 50% disk storage usage).

*Disk IOPS utilization:* The wa% is 2%, which means there are reads/writes to the disks.

*Load on the cluster:* The NLA is only 0.5, which isn't high.

## Options for Reducing Costs

### Scaling Down

The option of scaling down the cluster isn't feasible due to potential CPU saturation. The reason is that the only way in AWS to scale down is to use instances with 24 cores. With half of the current cores, the brokers will have CPU utilization of 80-85 percent, which might sometimes lead to a load average of more than 1.0 and to latency in the clients.

### Scaling In

Scaling in the cluster involves reducing it from 12 brokers to 8. This reduction is expected to affect your CPU and storage utilization as follows: CPU usage may rise from 40 to 60 percent, although it's worth noting that CPU utilization doesn't always scale linearly. The storage usage is expected to go from 50 to 75 percent. By implementing this change, you can anticipate a 30 percent savings on hardware costs without encountering any CPU or storage limitations. To accomplish this transition, you need to reassign the partitions across all the topics and adjust the number of partitions to ensure an even distribution among the remaining brokers.

## Example 4

Table 12-4 shows a Kafka cluster with ten brokers.

***Table 12-4.*** *A Kafka Cluster with Ten Brokers*

| # CPU Cores Per Broker | RAM Per Broker | Disk Storage Per Broker (NVMe) | %us | %sy | %wa |
|---|---|---|---|---|---|
| 16 | 122GB | 3.8TB (2 disks) | 22% | 12% | 1% |

| Total CPU% Utilization | Normalized Load Average | Used Disk Space in /var/lib/kafka | Network Processor Idle |
|---|---|---|---|
| 35% | 0.4 | 75% | 85% |

# Findings

*CPU utilization:* The cluster is over-provisioned in terms of CPU (only 35% CPU utilization).

*Disk storage usage:* The cluster uses 75 percent of its disk storage, which currently is enough.

*Disk IOPS utilization:* The `wa%` is 1 percent, which means there are almost no reads/writes to the disks.

*Load on the cluster:* The NLA is only 0.4, which isn't high.

*Network processor idle:* This is lower than expected, which shows there's either some bottleneck in the cluster or that there are not enough threads.

## Options for Reducing Costs

### Scale Down

To scale down the brokers, you'll need to use instances with eight cores, 16GB RAM, and one disk with 1.9TB storage. Let's see if these are enough:

- *CPU:* The cluster currently uses 35 percent of the CPU, so with half of the cores, the CPU utilization is expected to reach 70 percent, which is okay.
- *RAM:* There are very few reads from the disks, so all you can tell is that the current amount of RAM is sufficient for the brokers in order to avoid almost completely accessing the disks in order to read data that doesn't exist in the page cache. However, you can't tell whether with half of the current RAM, the brokers won't need to access the disks. The only way to know this is to scale down the cluster and monitor the CPU `wa%` (using the `top` command), the disk utilization (using the `iostat` command), and the misses from the page cache (using the `cachestat` script). These metrics will give you an indication whether the page cache has enough RAM in order to prevent the brokers from accessing the disks in order to serve the fetch requests of the consumers and other brokers (as part of the replication process).
- *Disk storage:* Since the current storage usage is already 75 percent, cutting the disk storage by half will leave the cluster with less storage capacity than required. So if the cluster will be scaled-down, you'll need to attach more disks to this broker, which is possible. In terms of disk storage and utilization, everything will remain the same since you'll keep the same amount of storage and IOPS as before.
- *Load on the cluster:* The normalized load average on the cluster is expected to be around 0.8, since the current load is 0.4 and you're going to reduce the cores by half, and it seems that most of the load originates from CPU us% utilization. Such a load is okay for Kafka clusters, but the cluster shouldn't reach a higher load than that.

### Scale In

To scale in the cluster, you can remove five brokers so that you'll be left with five brokers, each with the same number of cores, RAM, disk storage, and IOPS. The previous arguments for CPU, RAM, disk storage, and load on the cluster apply here. In order to keep the same amount of disk storage, you'll need to attach two more disks per broker, which is also possible.

## Recommendation

You can either scale down the cluster by half or scale in the cluster by half, but in both cases, you'll need to attach more disks in order to have the same amount of storage.

However, the issue of the low network processor idle% makes the decision whether to reduce resources from the cluster a tough one. This metric indicates the percentage of time that the network processor threads in a Kafka broker were idle and not processing any incoming requests from clients. A value of 85 percent means that for 15 percent of the time, the network threads were busy, and my experience shows that clients of Kafka clusters with such busy network threads usually experience some latency.

Although from the perspective of CPU, RAM utilization, and load on the cluster, it seems that you could reduce half of the cores, RAM, and hardware costs of the cluster, the low network threads idle% metric indicates there's a bottleneck that could cause clients of the clusters even greater latency.

That's why in the case of this cluster, it's better to check whether the clients suffer from latency rather than trying to scale down or scale in the cluster.

### Example 5

Table 12-5 shows a Kafka cluster with eight brokers.

***Table 12-5.** A Kafka Cluster with Eight Brokers*

| # CPU Cores Per Broker | RAM Per Broker | Disk Storage Per Broker (NVMe) | %us | %sy | %wa |
|---|---|---|---|---|---|
| 24 | 192GB | 15TB (2 disks) | 14% | 5% | 1% |

| Total CPU% Utilization | Normalized Load Average | Used Disk Space in /var/lib/kafka | Network Processor Idle |
|---|---|---|---|
| 20% | 0.25 | 40% | 99% |

## Findings

*CPU utilization:* The cluster is over-provisioned in terms of CPU (only 20% CPU utilization).

*Disk storage usage:* The cluster uses 40 percent of its disk storage, which currently is enough.

*Disk IOPS utilization:* The `wa%` is 1 percent, which means there are almost no reads/writes to the disks.

*Load on the cluster:* The NLA is only 0.25, which isn't high.

## Options for Reducing Costs

This is a pretty simple example of a cluster that can use brokers with half the cores, RAM, and storage. It can be achieved by either scaling down the cluster and using eight brokers with half the cores, RAM and storage, or by scaling in the cluster and using four brokers with the same instance type instead of eight brokers.

### Example 6

Table 12-6 shows a Kafka cluster with six brokers.

*Table 12-6. A Kafka Cluster with Six Brokers*

| # CPU Cores Per Broker | RAM Per Broker | Disk Storage Per Broker (NVMe) | %us | %sy | %wa |
|---|---|---|---|---|---|
| 12 | 96GB | 7.5TB | 20% | 7% | 1% |

| Total CPU% Utilization | Normalized Load Average | Used Disk Space in /var/lib/kafka | Network Processor Idle |
|---|---|---|---|
| 30% | 0.35 | 50% | 99% |

## Findings

*CPU utilization:* The cluster is over-provisioned in terms of CPU (only 30% CPU utilization).

*Disk storage usage:* The cluster uses only 50 percent of its disk storage.

*Disk IOPS utilization:* The `wa%` is 1 percent, which means there are almost no reads/writes to the disks.

*Load on the cluster:* The NLA is only 0.35, which isn't high.

## Options for Reducing Costs

### Scale Down

In AWS, you can use a lower instance type from the same instance family with eight cores, 64GB RAM, and 5TB storage (composed of two 2.5GB disks). This should increase the CPU utilization of the cluster to ±60 percent, the normalized load average to ±0.7, and the disk usage to 75 percent.

This move will reduce the cost of the cluster's hardware by 33 percent without reaching any CPU or storage bottleneck.

### Scale In

By scaling down the cluster from 12 brokers to 8, you can anticipate the CPU usage to increase from 40 to 60 percent and storage usage to rise from 50 to 75 percent. This adjustment is projected to lead to a 33 percent reduction in hardware costs without hitting any CPU or storage constraints. To facilitate this change, it's essential to reassign the partitions across all topics in the cluster and adjust the number of partitions to ensure a balanced distribution among the remaining brokers.

## Recommendation

Scaling in the cluster would reduce the hardware costs by 30 percent, since all the brokers will be replaced with brokers at 2/3 the price (and 2/3 the hardware resources). This will require migrating all the topics to the new brokers without having to change the number of partitions.

Scaling the cluster down would also reduce the costs by 33 percent and will require a reassignment of the partitions. It will also require a change in the number of partitions of all the topics in which their number of partitions doesn't divide equally by 8.

In this case, there's no better or worse approach, since both reduce the cost of the cluster equally.

# Summary

Scaling and optimizing Kafka clusters is a multifaceted challenge that requires a comprehensive understanding of both technical intricacies and financial considerations. This chapter delved into the various strategies for handling different scaling requirements, focusing on both cloud-based and on-premises Kafka clusters, and emphasized the importance of avoiding over-provisioning to reduce costs.

We began by exploring the constraints on resources such as RAM, CPU cores, and disk storage, and provided potential solutions, weighing their respective merits and challenges. While the cloud-based environment often affords more flexibility with options to scale up or out, on-premises clusters call for a meticulous evaluation of existing hardware and careful alignment with financial and managerial objectives. The importance of optimizing cost savings without sacrificing performance was highlighted, underscoring the need for accurate evaluation of actual workloads.

Furthermore, we offered insights into relevant Kafka metrics and OS considerations for assessing cluster usage, and provided six examples of over-provisioned clusters, illustrating how scaling in and scaling down of these clusters can be implemented. These practical examples elucidated the optimal scaling options to maximize cost reduction while maintaining stability.

It's important to emphasize that scaling and cost reduction of Kafka clusters are not merely technical endeavors, but are strategic undertakings that necessitate thoughtful planning and a well-rounded understanding of various influencing factors.

# Index

## A



## B



## C






E. Eldor, *Kafka Troubleshooting in Production*, https://doi.org/10.1007/978-1-4842-9490-1

## D



## E



## F



## G



## H

## I



## J, K

## L



## M



## N



## O



## P, Q

## R

## S



## T

## U, V, W, X, Y



## Z

Printed in the United States
by Baker & Taylor Publisher Services